A QUICK GUIDE TO METABOLIC DISEASE TESTING INTERPRETATION

A QUICK GUIDE TO METABOLIC DISEASE TESTING INTERPRETATION

Testing for Inborn Errors of Metabolism

Second Edition

PATRICIA JONES
Department of Pathology
UT Southwestern Medical Center
and Children's Health System of Texas

KHUSHBU PATEL
Department of Pathology
UT Southwestern Medical Center
and Children's Health System of Texas

DINESH RAKHEJA
Department of Pathology
UT Southwestern Medical Center
and Children's Health System of Texas

ACADEMIC PRESS

An imprint of Elsevier

Academic Press is an imprint of Elsevier
125 London Wall, London EC2Y 5AS, United Kingdom
525 B Street, Suite 1650, San Diego, CA 92101, United States
50 Hampshire Street, 5th Floor, Cambridge, MA 02139, United States
The Boulevard, Langford Lane, Kidlington, Oxford OX5 1GB, United Kingdom

Notices
Knowledge and best practice in this field are constantly changing. As new research and experience
broaden our understanding, changes in research methods, professional practices, or medical treatment may become necessary.

Practitioners and researchers must always rely on their own experience and knowledge in evaluating
and using any information, methods, compounds, or experiments described herein. In using such
information or methods they should be mindful of their own safety and the safety of others, including parties for whom they have a professional responsibility.

To the fullest extent of the law, neither the Publisher nor the authors, contributors, or editors, as-
sume any liability for any injury and/or damage to persons or property as a matter of products liability, negligence or otherwise, or from any use or operation of any methods, products, instructions, or
ideas contained in the material herein.

Library of Congress Cataloging-in-Publication Data
A catalog record for this book is available from the Library of Congress

British Library Cataloguing-in-Publication Data
A catalogue record for this book is available from the British Library

ISBN: 978-0-12-816926-1

For information on all Academic Press publications
visit our website at https://www.elsevier.com/books-and-journals

Publisher: Stacy Masucci
Editorial Project Manager: Megan Ashdown
Production Project Manager: Omer Mukthar
Designer: Greg Harris

Typeset by Thomson Digital

Working together
to grow libraries in
developing countries

www.elsevier.com • www.bookaid.org

Contents

Preface

The timely and accurate diagnosis of inborn errors of metabolism (IEM) is a vitally important aspect of healthcare, and especially of pediatric laboratory medicine. Rapid diagnosis leads to rapid treatment onset that improves the long-term prognosis of many IEM, positively impacting the patient's life, health, and continued growth and development. Testing for IEM is complex and usually involves relatively esoteric analysis using laboratory-developed tests, including organic acid, amino acid, and acylcarnitine analysis. In addition to the testing itself, accurate interpretation of the results of these tests requires a significant amount of practice to perform adequately. Pattern recognition is an important factor and a skill that can be gained through time and effort. Some IEM are so rare that they might be seen only once or twice in a practitioner's lifetime, and may occasionally be difficult to recognize. However, there are also "common" disorders of metabolism that can be easily recognized by looking for common metabolites in specific patterns.

Residents and fellows observing the interpretation of these tests, or learning to interpret them, often request literature references for help in recognizing the best testing options to proceed with, as well as recognizing test result patterns. Information that has been found to be especially helpful includes example organic acid chromatograms demonstrating specific disease-related metabolites for the various IEM, as well as information on patterns of test results from organic acids, amino acids, and acylcarnitines for the specific IEM. This second edition of the Quick Guide is intended as such a reference for the IEM that are commonly seen in clinical practice. It provides information to aid in test utilization and interpretation for the diagnosis of specific IEM.

Acknowledgment

The authors of this Quick Guide would like to express their gratitude to the technologists who work in the lab and who pulled representative chromatographs for this work: Vivian Jones, Susan Fisher, Mashaer Sunbul, and Thomas Duong.

SECTION 1

Introduction

CHAPTER 1

Introduction

Inborn errors of metabolism (IEMs) are genetic disorders that result from defects in energy production and/or the metabolism of macromolecules. Individually, IEMs are rare diseases; however, collectively they are quite common with an incidence of approximately 1 in 2500 births. These disorders often present with a range of clinical phenotypes, most frequently with non-specific presentations ranging from poor feeding, vomiting, temperature instability and lethargy to seizures and coma. While newborn screening programs identify infants at risk for many of these disorders, further laboratory testing is essential for confirmation and diagnosis as well as identifying those IEMs that are not covered by the screening programs. Specialized testing for confirmation includes of amino acid, acylcarnitine and urine organic acid analysis and is typically performed at either reference laboratories or biochemical genetics laboratories within pediatric hospitals. The methodologies used for performing these tests are described below and subsequent chapters that follow highlight commonly seen disorders and the application of these techniques in their diagnosis. Thus, this Quick Guide is intended to serve as a reference for example laboratory findings commonly observed in IEMs that are commonly encountered in clinical practice. It is not meant to be an inclusive compendium of all IEM that can be diagnosed via these methodologies, nor is it mean to be comprehensive with respect to the details of IEM etiology and clinical picture. The guide provides a pathway diagram for each IEM delineating the metabolite(s) that would be seen in the disorder. One or more representative chromatographs from each of the included disorders are presented when available. Other laboratory testing or metabolite markers and the expected results that would be informative in the diagnosis are also detailed.

1 Methodologies

Here we describe the most common methods that are used by clinical laboratories for the diagnosis of IEMs. Most of the described methodologies are high-complexity laboratory developed tests. Though methods can vary among laboratories, HPLC/UPLC, GC–MS and LC-MS/

MS are the most commonly used techniques. Over 50 different disorders can be identified primarily through plasma amino acid and acylcarnitine analysis alone, while many more IEMs can be uncovered through urine organic acid analysis as this methodology can identify several hundred different types of organic acids. However, given the rarity of individual disorders and the complexity of preanalytical and analytical factors that impact testing, interpretation of results is still often challenging. Careful recognition of clinical features and integration of the multiple biochemical profiles with other laboratory testing is essential for the investigation and diagnosis of IEMs. Therefore, providing an accompanying interpretation that incorporates both the clinical and biochemical findings to the clinical team is necessary due to the complexity of patterns that are observed routinely.

1.1 Amino acid analysis

Since the 1950s, ion-exchange chromatography with post-column ninhydrin derivatization has been the recognized "gold standard" for amino acid analysis. Other chromatographic-based methods that utilize pre-column derivatization have also been described and used clinically. More recently, alternative approaches using LC-MS/MS are gaining popularity. One limitation to be aware of when interpreting results from methods that rely solely on chromatographic separation is the interference of co-eluting compounds, which can be in the form of endogenous or exogenous substances. LC-MS/MS offers a significant advantage with analytical specificity; however, techniques for sample preparation and derivatization for mass spectrometry analysis are typically more cumbersome when compared to sample preparation for chromatographic techniques.

Levels of amino acids are relatively constant in the blood, and ideally fasting samples are preferred for the diagnosis of defects in amino acid metabolism. However, this is challenging in newborns and most samples sent to the laboratory in daily practice are typically collected in the non-fasting state. This often results in mild non-specific elevations of several amino acids. Additionally, hemolysis can increase glutamate, aspartate and taurine levels, which are present in higher concentrations in red blood cells (RBC) compared to plasma. Hemolysis can also result in increased ornithine concentrations as arginase released from RBC lysis can convert arginine to ornithine. Plasma samples will also degrade with time, resulting in glutamine concentrations decreasing and glutamate rising as glutamine is converted to glutamate.

While plasma is the preferred specimen for identification of most amino acid disorders, in some cases, urine or CSF may be preferred or useful in addition to plasma. Urine amino acid analysis can aid the diagnosis of disorders of amino acid transport such as lysinuric protein intolerance, cystinuria and Hartnup disease. Glycine measurement in CSF is useful in diagnosis of glycine encephalopathy. Concentrations of amino acids in urine are impacted by diet, diurnal variation, and urinary tract disease; therefore, caution should be exercised when interpreting these results. Additionally, CSF should be free of blood contamination due to differences in the concentration of amino acids between plasma and CSF. Table 1.1 lists amino acid alterations seen with IEM as well as from non-IEM sources.

Table 1.1 Amino acids in inborn errors of metabolism (IEM) and other sources.

Amino acid (increased)	Associated IEM	Non-IEM sources
Alloisoleucine	Maple Syrup Urine Disease (MSUD)	
Arginine	Argininemia (low in other urea cycle defects and lysinuric protein intolerance)	
Argininosuccinic Acid	Argininosuccinate Lysase Deficiency	
Isoleucine	MSUD	Diet
Leucine	MSUD	Diet
Valine	MSUD	Diet
Citrulline	Citrullinemia Citrin Deficiency Argininosuccinic aciduria	
Glutamine	Urea Cycle Defects	Liver dysfunction
Glycine	Glycine enecephalopathy Methymalonic aciduria Propionic Acidemia	
Methionine	Homocystinuria Tyrosinemia Type 1	Liver dysfunction
Phenylalanine	Phenylketonuria	Liver dysfunction
Tyrosine	Tyrosinemia Types 1-3	Liver dysfunction
Glycine (CSF)	Glycine encephalopathy	
Arginine (urine)	Lysinuric protein intolerance	
Lysine (urine)	Lysinuric protein intolerance	
Ornithine (urine)	Lysinuric protein intolerance	
Neutral amino acids (urine)	Hartnup Disease	

1.2 Acylcarnitine analysis

Tandem mass spectrometry (MS/MS) is the method most commonly employed by clinical laboratories for the analysis of carnitine esters. The sample is typically derivatized with n-butanolic HCl. The resultant acylcarnitine species are identified as n–butyl–esters using electrospray ionization MS/MS. Chromatographic separation can be performed if separation of isomers is necessary; however, this is rarely necessary when other testing results from urine organic acid or acylglycine analysis are available. Though the quantitative result of individual acylcarnitine species are reported, an interpretation of results is usually included as part of the analysis. Through an acylcarnitine profile, fatty acid oxidation disorders as well as organic acid disorders can be detected.

Plasma or serum is the preferred specimen for detection of fatty acid oxidation or organic acid defects through acylcarnitine analysis. For detection of most fatty acid oxidation defects, the sample should be collected at the presentation of symptoms. Urine acylcarnitine analysis is less useful except in the cases where glutaric acidemia type I is suspected. Individual abnormalities in acylcarnitine species are analyzed against validated reference intervals, which need to be established by the laboratory and validated periodically. Modest abnormalities in plasma acylcarnitines need to be correlated with the patient's carnitine levels. Several drugs (antibiotics) and food additives can lead to elevations in one or more acylcarnitine species. Dietary intake of fatty acids (ketogenic diets or medium chain triglycerides) can also result in artifactual increase in acylcarnitine species. Acylcarnitine abnormalities associated with IEMs and non-IEM sources are provided in Table 1.2.

1.3 Organic acid analysis

GC-MS is the most widely used method for organic acid analysis. The technique involves extraction of organic acids with ethyl acetate, acidification with HCl, followed by trimethylsilyl (TMS) derivatization, which allows for creation of characteristic fragments upon electron impact as these compounds leave the GC column and enter the mass spectrometer. The resulting mass spectrum from these fragments is used for identification of compounds through library searches for spectral matches. The data from GC–MS analysis is analyzed in the form of a total ion chromatogram (TIC) for qualitative interpretations. Reporting and analysis strategies differ among laboratories, where some laboratories may offer metabolite quantification relative to creatinine concentration and some laboratories may

Table 1.2 Acylcarnitines in associated IEM and non-IEM sources.

Acylcarnitine	Associated IEM	Non-IEM sources
C0 (decreased)	Carnitine uptake defect (secondary carnitine deficiencies)	Prematurity and infant of vegan mom
C0	CPT1 deficiency (with C16, C18 decreased)	Sepsis, carnitine supplementation
C3	Propionic Acidemia, Methylmalonic Acidemia, Biotinidase deficiency	Branched-chain intermediate, gut bacteria, prematurity, low vitmain B12
C4	SCAD deficiency, Multiple acyl-CoA dehydrogenase deficiency	Branched-chain intermediate
C5	Isovaleric acidemia, 3-Methylbutyryl-CoA dehydrogenase deficiency	Valproate, antibiotics, branched-chain intermediate
C4-OH	SCHAD deficiency	
C5-OH	3-Methylcrotonyl-CoA carboxylase deficiency, Holocarboxylase synthetase deficiency, HMG-CoA lyase deficiency, Biotinidase deficiency, 3-Methylglutaconyl-CoA hydratase deficiency, 2-Methyl-3-hydroxybutyryl-CoA dehydrogenase deficiency	Prematurity
C8	MCAD deficiency (C6, C10, C10:1 also increased)	MCT (C6 and C10 also increased)
C3DC	Malonyl-CoA carboxylase deficiency	
C4DC	Succinyl-CoA synthetase deficiency	
C5DC	Glutaric acidemia type I	Kidney disease
C10-OH	M/SCHAD deficiency, MCKAT deficiency	
C14:1	VLCAD deficiency (C14, C14:2 also increased)	Ketosis
C16	CPTII deficiency C14, C14:2 also increased), carnitine/acylcarnitine translocase (CACT) deficiency (C18:2, C18:1, C18 also increased)	

(*Continued*)

Table 1.2 Acylcarnitines in associated IEM and non-IEM sources. (*Cont.*)

Acylcarnitine	Associated IEM	Non-IEM sources
C16-OH	LCHAD deficiency (C16:1-OH, C18:1-OH, C18-OH also increased), Trifunctional protein (TFP) deficiency (with C16:1-OH, C18:1-OH, C18-OH also increased)	Sepsis, ketosis. C16:1-OH can be increased from cephalosporin

Source: Adopted from Rinaldo P, Cowan TM, Matern D. Acylcarnitine profile analysis. Genet Med 2008;10:151–6.

offer a qualitative assessment of metabolites. Nevertheless, interpretation of the organic acid pattern is necessary due to the number and complexity of metabolites and patterns that can be present.

Organic acids are low molecular weight, highly water-soluble compounds, which are primarily excreted in the urine. Therefore, urine is the preferred specimen for organic acid analysis. Organic acid profiling using plasma or serum has little to no utility and should be not be used in the investigation of IEMs. A random urine sample collected without preservatives is acceptable for analysis. A random urine sample only provides the metabolite profile of a snapshot in time. A urine sample collected at the time of metabolic decompensation (presentation of symptoms) is sometimes necessary to detect the diagnostic metabolite(s). Additionally, some diagnostically important compounds are labile or volatile; therefore, appropriate storage and prompt analysis is essential. Furthermore, several preanalytical factors, such as diet, fasting and hydration status, and drug treatment regimens, can confound interpretation. These factors must be considered during interpretation and diagnosis should always be made with these limitations in mind. Table 1.3 provides a list of metabolites commonly seen in urine, their associated IEMs as well as their non–IEM sources.

Table 1.3 Urinary organic acids with associated IEM and non-IEM sources.

Acid/Metabolite	Associated IEMs	Non-IEM sources
Phenylacetate	PKU; BH4 deficiency	Intestinal bacterial origin (from phenylalanine)
Phenyllactate	PKU; BH4 deficiency	Bacterial gut metabolism (D-form); liver diseases
Phenylpyruvate	PKU; BH4 deficiency	Bacterial gut metabolism; liver diseases
2-Hydroxyphenylacetate	PKU; BH4 deficiency	Uremia
Mandelate	PKU	Preservative in albumin solution for intravenous perfusion; methenamine mandelate; gastrointestinal malabsorption diseases
4-Hydroxyphenylacetate	Tyrosinemia; PKU; hawkinsinuria	Bacterial gut metabolism and bacterial contamination (from tyrosine); short bowel syndrome; liver diseases
4-Hydroxyphenyllactate	Tyrosinemia; PKU; hawkinsinuria	Bacterial gut metabolism; short bowel syndrome; liver diseases (e.g., secondary to PA, galactosemia, fructosemia); scurvy; lactic acidosis
4-Hydroxyphenylpyruvate	Tyrosinemia; hawkinsinuria	VPA; liver diseases (e.g., secondary to PA, galactosemia, fructosemia)
4-Hydroxycyclohexylacetate	Hawkinsinuria	Bacterial gut metabolism (?)
Succinylacetoacetate	Tyrosinemia type I	
Succinylacetone	Tyrosinemia type I	
N-Acetyltyrosine	Tyrosinemia	Some parenteral solutions
Homogentisate	Alcaptonuria	
3-Keto-2-methylvalerate	PA; MMA; β-ketothiolase deficiency	
Methylcitrate	PA; MMA; multiple carboxylase deficiency	Malnutrition
Propionylglycine	PA; MMA	
Tiglylglycine	PA; 2-methyl-3-hydroxybutyryl-CoA DH deficiency; multiple carboxylase deficiency; respiratory chain defects	Reye & Reye-like syndromes; VPA

(Continued)

Table 1.3 Urinary organic acids with associated IEM and non-IEM sources. (*Cont.*)

Acid/Metabolite	Associated IEMs	Non-IEM sources
3-Hydroxypropionate	PA; MMA; multiple carboxylase deficiency; succinic semialdehyde DH deficiency; methylmalonic semialdehyde DH deficiency; lactic acidosis (with pyruvate carboxylase deficiency)	Bacterial metabolism and contamination; short bowel syndrome; lactic acidosis
Methylmalonate	MMA; transcobalamine II deficiency; malonic aciduria	B12 vitamin deficiency, pernicious anemia; bacterial gut metabolism; gastroenteritis in very young infants; short bowel syndrome; apnea; "benign" MMA; decreased GFR (in plasma); malnutrition
3-Hydroxy-3-methylglutarate	HMG–CoA lyase deficiency	Ketosis
Glutaconate	GA I	
3-Hydroxyglutarate	GA I	
Glutarate	GA I; MAD deficiency (GA II); 2-amino/2-ketoadipic aciduria; malonic aciduria; other mitochondrial dysfunctions	2-Ketoglutarate degradation; bacterial gut metabolism; uremia; ethylene glycol poisoning; lithium
2-Hydroxyglutarate (L– or D– form)	2-Hydroxyglutaric acidurias; MAD deficiency (GA II); malonic aciduria	Bacterial contamination (D-form); lithium; uremia; increase with younger age; 2-ketoglutarate degradation
N-Acetylaspartate	Canavan disease	
Fumarate	Fumarase deficiency; respiratory chain defects; pyruvate carboxylase deficiency; pyruvate dehydrogenase complex (E1, E3) deficiency	Lithium; renal tubular reabsorption defect (fumaric aciduria); increase with younger age

Compound	Deficiencies/defects	Other causes
Succinate	Malonic aciduria; fumarase deficiency; respiratory chain defects; pyruvate carboxylase deficiency; pyruvate dehydrogenase complex (E1, E3) deficiency	Bacterial (on storage); 2-ketoglutarate degradation; lithium; ketosis; tissue ischemia; increase with younger age
Malate	Respiratory chain defects; pyruvate carboxylase deficiency; pyruvate dehydrogenase complex deficiencies	Lithium; uremia; increase with younger age
Lactate and pyruvate	Pyruvate dehydrogenase complex deficiencies; oxidative phosphorylation and respiratory chain defects, Krebs acid cycle defects, gluconeogenesis defects; MAD deficiency (GA II);VLCAD deficiency; multiple carboxylase deficiency; citrullinemia, glycerol kinase deficiency; HMG–CoA lyase deficiency	Gut bacteria and bacterial contamination (D-lactate); short bowel syndrome (D-lactate); secondary lactic acidosis (e.g., apnea, septicemia, seizures, respiratory or cardiac insufficiency); diabetic ketoacidosis; Reye & Reye-like syndromes; increase with younger age; saccharose, fructose, lactose; drugs inducing hyperlactemia; dialysis bath; MCT administration
Glycerol	Gylcerol kinase deficiency; fructose-1,6-phosphatase deficiency	Contamination (suppository, emollients); uremia
2-Hydroxy-3-methylvalerate	MSUD	Short bowel syndrome (D-form)
2-Hydroxyisocaproate	MSUD	Ketosis; lactic acidosis
2-Hydroxyisovalerate	MSUD	Lactic acidosis; ketosis
2-Keto-3-methylvalerate	MSUD	Lactic acidosis; ketosis
2-Ketoisocaproate	MSUD	Lactic acidosis; ketosis
2-Ketoisovalerate	MSUD	Lactic acidosis; ketosis
4-Hydroxyisovalerate	IVA	
Isovalerylglycine	IVA; MAD deficiency (GA II); ethylmalonic aciduria	Valproate

(Continued)

Table 1.3 Urinary organic acids with associated IEM and non-IEM sources. *(Cont.)*

Acid/Metabolite	Associated IEMs	Non-IEM sources
3-Hydroxyisovalerate	IVA; multiple carboxylase deficiency; HMG–CoA lyase deficiency; 3-methylcrotonyl–CoA carboxylase deficiency; 3-methylglutaconyl–CoA hydratase deficiency; succinyl-CoA:3-oxoacid-CoA transferase deficiency; MAD deficiency (GA II)	Reye & Reye-like syndromes; VPA; ketosis
Isobutyrylglycine	MAD deficiency (GA II); ethylmalonic aciduria	
2-Methylbutyrylglycine	MAD deficiency (GA II); ethylmalonic aciduria	VPA
Ethylmalonate	SCAD deficiency; MAD deficiency (GA II); acetyl-CoA carboxylase deficiency; ethylmalonic aciduria	Jamaican vomiting sickness; neonates on fasting; diet (?)
Butyrylglycine	SCAD deficiency; MAD deficiency (GA II); ethylmalonic aciduria	MCT administration; ketosis; Jamaican vomiting sickness
Malonate	Malonyl-CoA–decarboxylase deficiency	
Thymine	Dihydropyrimidine DH deficiency	Caffeine (?)
Uracil	Dihydropyrimidine DH deficiency; OTC deficiency; citrullinemia	Caffeine (?)

Compound	Metabolic/disease	Other sources
Pyroglutamate	Glutathione synthetase deficiency; hawkinsinuria; homocystinuria; OTC deficiency; PA	From glutamine of hydrolyzed proteins (infant formula); acetaminophen; vigabatrin; fludoxacillin, netilmicin (?); glutamine degradation (in hyperammonemia, urea cycle defects); vegetarian or low-protein diets, undernutrition; iron oxoprolinate; Steven–Johnson syndrome; burns; premature newborns; transitory (?); glycine deficiency; increase with younger age; renal insufficiency; pregnancy (increased metabolic demand for glycine)
Orotate	Argininemia; citrullinemia; OTC deficiency; hyperornithinemia-hyperammonemia-homocitrullinuria syndrome; lysinuric protein intolerance; purine nucleoside phosphorylase deficiency; Lesh–Nyhan disease	Allopurinol treatment; azauridine; high cell turnover (tissue breakdown, menstruation); folate malabsorption
3,4-Dihydroxybutyrate	Succinic semialdehyde DH deficiency	Diet
4,5-Dihydroxyhexanoate	Succinic semialdehyde DH deficiency	
4-Hydroxybutyrate	Succinic semialdehyde DH deficiency	
Dicarboxylic acids (adipate, suberate, sebatate)	Fatty acid oxidation defects (SCAD, SCHAD, MCAD, VLCAD, LCHAD); CPT II Deficiency; peroxisomal disorders	Ketosis (?)
Ketones (3-Hydroxybutyrate, acetoacetate)	Lactic acidosis; respiratory chain defects; diminished in medium and long chain fatty acid oxidation defects	Seriously ill states: infection, malnutrition, fever, seizures, liver diseases, pulmonary stenosis; MCT administration; ketosis; VPA or acetaminophen; lactic acidosis; hypoglycemia; Reye & Reye–like syndromes; Jamaican vomiting sickness

Source: Adopted from Kumps A, Duez P, Mardens Y. Metabolic, nutritional, iatrogenic, and artifactual sources of urinary organic acids: a comprehensive table. Clin Chem 2002;48(5): 708–17.

Further reading

[1] Phipps WS, Jones PM, Patel K. Amino and organic acid analysis: Essential tools in the diagnosis of inborn errors of metabolism. In: Makowski G, editor. Advances in Clinical Chemistry. St. Louis: Elsevier; 2019. Chapter 2, pp 59-104.
[2] Rinald P, Cowan TM, Matern D. Acylcarnitine profile analysis. Genet Med 2008;10:151–6.
[3] Jones PM, Bennett MJ. Urine organic acid analysis for inherited metabolic disease by gas chromatography-mass spectrometry. In: Garg U, Hammett-Stabler CA, editors. Methods in molecular biology. New York: Humana Press; 2009. p. 423–31. Chapter 41.
[4] Hong, P. et al. Compilation of amino acids, drugs, metabolites and other compounds in masstrak amino acid analysis solution. Waters Corporation, Application notes. 2019
[5] Thompson JA, Miles BS, Fennessey PV. Urinary organic acids quantitated by age groups in a healthy pediatric population. Clin Chem 1977;23:1734–8.
[6] Kumps A, Duez P, Mardenas Y. Metabolic, nutritional, iatrogenic, and artifactual sources of urinary organic acids: a comprehensive table. Clin Chem 2002;48(5):708–17.

Organic acidurias

CHAPTER 2

Disorder: Glutaric acidemia type 1

1 Synonyms

Glutaric aciduria, Glutaryl-CoA dehydrogenase deficiency

2 Brief synopsis

2.1 Incidence

~1:30,000–40,000 newborns; ~1:300 in Amish and Canadian Ojibwa.

2.2 Etiology

GA1 (OMIM #231670) is a disorder of the catabolic pathways of the amino acids L-tryptophan, L-lysine, and L-hydroxylysine. Glutaryl-CoA dehydrogenase (EC1.3.99.7) is the enzyme that catalyzes the conversion of the α-ketoacid form of these amino acids through glutaryl-CoA and glutaconyl-CoA to its crotonyl-CoA form. When activity is deficient and the conversion does not proceed as usual, intermediate metabolites accumulate, including glutaric acid, 3-hydroxyglutaric acid, glutaconic acid, and glutarylcarnitine (Fig. 2.1). These intermediates are believed to damage the brain, especially the basal ganglia. Mutations in the *GCDH* gene cause GA1, preventing production of glutaryl-CoA dehydrogenase or resulting in the production of a defective enzyme. The gene locus for this enzyme is 19p13.2, and the disorder is inherited in an autosomal recessive fashion. The gene encodes a protein of 438 amino acids, and more than 100 disease-causing mutations have been identified. Because no mutation is prominent, most GA1 patients are heterozygous for two different mutations. There is not a good genotype-phenotype correlation in GA1. Disease outcome and prognosis are more dependent on early diagnosis and appropriate treatment of acute encephalopathic crises than on specific mutations or degree of deficiency.

3 Clinical presentation

Initially the physical signs are mild and often occur after a period of normal development. Symptoms rarely occur after 5 years of age. The disorder often presents as an acute encephalopathic episode, usually between 4 and

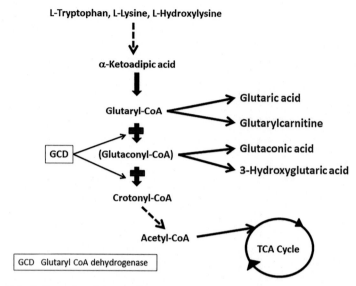

Figure 2.1 *Pathway involving arginase.*

18 months of age, which may be precipitated by common infections, fasting, routine immunizations, or minor trauma. Affected babies demonstrate loss of head control, hypotonia, and seizures. There may be macrocephaly and frontal bossing, a high arched palate, difficulty with feeding, and irritability. During acute crises, the basal ganglia are damaged, resulting in permanent disability. The clinical picture often progresses to a severe dystonic–dyskinetic disorder with athetoid movements. There is relative preservation of intellect. It is also possible for patients with GA1, even occasionally siblings of severely affected patients, to avoid developing neurological problems, and an adult onset form has been described that shows different degrees of leukoencephalopathic changes.

4 Diagnostic compounds

4.1 Urine organic acid profile

Glutaric acid and 3-hydroxyglutaric acid will almost always be present. 3-Hydroxyglutaric acid may be just barely detectable as some individuals with GA1 are considered "low excretors" and may show only slight, if any, excretion of these biomarkers. Glutaconic acid may or may not be present (Fig. 2.2).

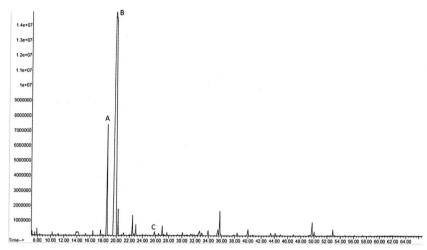

Figure 2.2 *Glutaric acidemia, type 1, classic.* (A) Internal standard, (B) glutaric acid, and (C) 3-hydroxyglutaric acid.

4.2 Acylcarnitine profile

Glutarylcarnitine (C5DC) is usually elevated in the serum, but not always. Glutarylcarnitine is essentially always elevated in the urine, and this is one of the few indications for a urinary acylcarnitine profile. Free carnitine is often very low. Glutarylcarnitine can non-specifically be elevated in the setting of kidney disease.

4.3 Amino acids

Not diagnostic for GA1.

4.4 Example chromatograph

4.5 Other important diagnostic/monitoring compounds

None applicable.

5 Newborn screening

Elevated glutarylcarnitine (C5DC) is the biomarker used to screen for GA1 on the newborn screening programs.

6 Follow-up/confirmatory testing

Molecular testing for the GCDH gene is available to help confirm the diagnosis.

7 Interferences and assay or interpretation quirks

Elevated excretion of glutaric acid without 3-hydroxyglutaric acid may indicate glutaric acidemia type 2 (GA2) if other compounds such as ethylmalonic acid and various glycines are present.

Non-IEM sources of glutaric acid include bacterial metabolism or contamination, 2-ketoglutarate degradation, uremia, and ethylene glycol poisoning.

Further reading

[1] Genetics Home Reference, NIH US National Library of Medicine. https://ghr.nlm.nih.gov/condition/glutaric-acidemia-type-i# accessed 4/5/2019.
[2] Hoffmann GF, Burlina A, Barshop BA. Organic acidurias. In: Sarafoglou K, Hoffman GF, Roth KS, editors. Pediatric endocrinology and inborn errors of metabolism. 2nd ed. New York: McGraw-Hill Co; 2017. p. 209–50.
[3] GeneReviews, NIH, Glutaric Acidemia, Type 1. https://www.ncbi.nlm.nih.gov/books/NBK546575/ accessed 12/31/2019.
[4] NIH Genetic and Rare Diseases Information Center. Glutaric Acidemia, type 1. https://rarediseases.info.nih.gov/diseases/6522/glutaric-acidemia-type-i accessed 12/31/2019.

CHAPTER 3

Disorder: Glutaric acidemia type 2

1 Synonyms

Multiple acyl-CoA dehydrogenase deficiency (MADD), electron transfer flavoprotein deficiency (ETFA, ETFB, ETFDH deficiencies), ethylmalonic-adipic aciduria.

2 Brief synopsis

2.1 Incidence

The incidence of glutaric acidemia type 2 (GA2) is unknown, although it is believed to be very rare.

2.2 Etiology

Glutaric academia type 2 (GA2) (OMIM #231680) results from defects of electron transfer from primary flavoprotein dehydrogenases to coenzyme Q10 in the mitochondrial electron transport chain. The electron transfer flavoprotein (ETF; composed of ETFα and ETFβ subunits) and the electron transfer flavoprotein dehydrogenase, (ETFDH, also known as ETF-ubiquinone oxidoreductase (ETF-QO)), are responsible for this electron transfer process. Therefore, mutations in any of the three genes that encode these proteins may give rise to GA2. The gene *ETFA*, encoding the ETFα subunit, resides on chromosome region 15q23-q25. The gene *ETFB*, encoding ETFβ subunit, resides on chromosome region 19q13.3. The gene *ETFDH*, encoding ETFDH, resides on chromosome region 4q32-qter. In GA2, the defective electron transfer affects multiple intramitochondrial flavoprotein dehydrogenases involved in fatty acid, amino acid, and choline metabolism. (Fig. 3.1) GA2 is considered a fatty acid oxidation disorder and/or included with the mitochondrial fatty acid oxidation defects. It is inherited in an autosomal recessive manner.

3 Clinical presentation

GA2 has three clinical presentations. The patients may have a neonatal onset of hypoglycemic encephalopathy with (type I) or without (type II) congenital anomalies, or they may have a later-onset, milder disease with progressive

A Quick Guide to Metabolic Disease Testing Interpretation
http://dx.doi.org/10.1016/B978-0-12-816926-1.00003-1

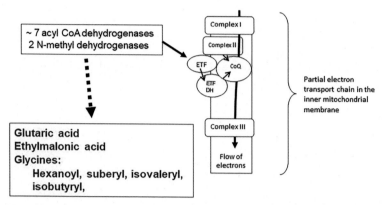

Figure 3.1 *Pathway involving electron transfer from primary flavoprotein dehydrogenases to coenzyme Q10 in the mitochondrial electron transport chain.*

muscle weakness (type III). The neonatal-onset disease (type I and type II) is characterized by fatal multiorgan failure including hepatic encephalopathy, cardiomyopathy, skeletal muscle weakness, renal tubular acidosis, hypoglycemia, and metabolic acidosis. A characteristic odor of sweaty feet is discernible in the body fluids. The type II presentation, which is milder and of later onset, may present with hepatic encephalopathy induced by stresses such as fasting, fever, or exercise. Later presentations may include isolated rhabdomyolysis. There is clear genotype-phenotype correlation with no measurable enzyme activity in type I disease, small residual enzyme activity in type II disease, and higher residual enzyme activity in type III disease. Type 1 GA2 is uniformly fatal, usually within the first few days of life. Treatment is difficult due to the many enzymes involved but includes dietary restriction of affected amino acids and long chain fatty acids, except those essential to proper development. High dose riboflavin may be helpful in milder cases. Prognosis is dependent on the amount of residual activity or flavoprotein present.

4 Diagnostic compounds

4.1 Urine organic acid profile

The metabolic defects associated with GA2 give rise to a characteristic urinary organic acid profile of elevated multiple short-chain aliphatic mono- and dicarboxylic acids and their glycine conjugates, including butyrylglycine, hexanoylglycine, suberylglycine, isovalerylglycine, isobutyrylglycine, 2-methylbutyrylglycine, and lactic, ethylmalonic and glutaric acids (Fig. 3.2). These compounds may be seen in milder cases only under conditions of metabolic decompensation and may not be present when the individual is metabolically stable.

4.2 Acylcarnitine profile

Acylcarnitine profile may show elevations in some or all of the following: C4-, C5-, C5DC-, C6-, C8-, C10-, and C12-carnitines. Low total carnitine may also be present.

4.3 Amino acids

Increased concentration of sarcosine may be present in the plasma.

4.4 Example chromatographs

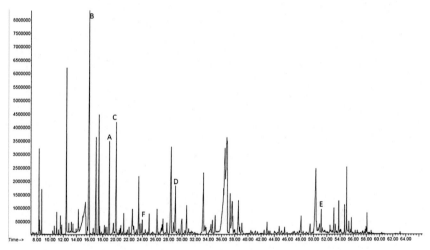

Figure 3.2 *Glutaric acidemia, type 2.* (A) Internal standard, (B) ethylmalonic acid, (C) glutaric acid, (D) hexanoylglycine, (E) suberylglycine, and (F) isovalerylglycine.

4.5 Other important diagnostic/monitoring compounds

Lactic acid is usually elevated. Like most other fatty acid oxidation defects, a nonketotic hypoglycemia is seen upon fasting.

5 Newborn screening

GA2 is not one of the core disorders in the Newborn Screening Recommended Uniform Screening Panel (RUSP), however it is one of the secondary conditions that are diagnosed in the course of working up the differential on the core conditions. The serum acylcarnitine analysis shows a profile of elevated C4- and C5- and various other acylcarnitines up to C12-carnitine.

6 Follow-up/confirmatory testing

Gene sequencing of *ETFA*, *ETFB*, and *ETFDH* may be performed for mutation detection.

7 Interferences and assay or interpretation quirks

In mild cases of GA2 when the individual is metabolically stable, the characteristic biomarkers may not be produced in enough quantities to be detectable by organic acid or acylcarnitine analysis. Both these analyses may be normal.

Isovalerylglycine and 2-methylbutyrlglycine may be present in the urine due to valproate. Ethylmalonic acid can be nonspecifically present in the urine in neonates during fasting.

Further reading

[1] Genetics Home Reference, NIH US National Library of Medicine. https://ghr.nlm.nih.gov/condition/glutaric-acidemia-type-ii# accessed 4/8/2019.
[2] Vockley J, Longo N, Andresen BS, Bennett MJ. Mitochondrial fatty acid oxidation defects. In: Sarafoglu K, Hoffmann GF, Roth KS, editors. Pediatric endocrinology and inborn errors of metabolism. 2nd ed. New York: McGraw-Hill Co; 2017. p. 125–44.
[3] NIH Genetic and Rare Diseases Information Center. Glutaric Acidemia, type II. https://rarediseases.info.nih.gov/diseases/6523/glutaric-acidemia-type-2 Accessed 12/31/2019.

CHAPTER 4

Disorder: 2-Hydroxyglutaric aciduria

1 Distinct disorders that fall under 2HGA

D-2-hydroxyglutaric aciduria (D2HGA) Types I (D2HGA1 OMIM # 600721) and II (D2HGA2 OMIM # 613657).

L-2-hydroxyglutaric aciduria (L2HGA OMIM# 236792).

Combined D-2- and L-2-hydroxyglutaric aciduria (D2L2AD OMIM # 615182).

2 Brief synopsis

2.1 Incidence

<1:200,000

2.2 Etiology

L-2-hydroxyglutarate (L2HG) (equivalent to S-2HG using the R/S nomenclature) and D-2-hydroxyglutarate (D2HG) (equivalent to R-2HG using the R/S nomenclature) are stereoisomers with identical physicochemical properties such as melting point and solubility. However, they have different three-dimensional spatial configurations that are mirror images of each other, and they rotate plane-polarized light either clockwise or counterclockwise. The different three-dimensional spatial configurations have biologic significance because the two compounds are recognized by different enzymes and are therefore associated with distinct metabolic disorders.

L2HG is formed from α-KG by the side activity of L-malate dehydrogenase, which normally catalyzes the interconversion of L-malate and oxaloacetate in the tricarboxylic acid (TCA) cycle. L2HG, which has no known metabolic function in eukaryotes, is converted back to α-KG by the catalytic action of L-2-hydroxyglutarate dehydrogenase (L2HGDH) (EC1.1.99.2), a mitochondrial FAD-linked dehydrogenase. A deficiency of L2HGDH, caused by mutations in the gene *L2HGDH* located on chromosome region 14q22.1, is responsible for the metabolic disorder L2HGA (OMIM # 236792).

A Quick Guide to Metabolic Disease Testing Interpretation
http://dx.doi.org/10.1016/B978-0-12-816926-1.00004-3

D2HG is formed by the activity of hydroxyacid-oxoacid transhy-drogenase (HOT). The catalyzation of gamma-hydroxybutyrate and α-ketoglutarate by HOT generates succinic semialdehyde and D2HG, re-spectively. D2HG is converted back to α-KG by the catalytic action of D-2-hydroxyglutarate dehydrogenase (D2HGDH), which is also FAD-linked. A deficiency of D2HGDH, caused by mutations in the gene *D2H-GDH* located on chromosome 2q37.3, give rise to the metabolic disorder D2HGA Type I (OMIM # 600721). A subset of patients with D2HGA (Type II) were later shown to carry heterozygous germline mutations in the *IDH2* gene, which codes for the enzyme isocitrate dehydrogenase 2 and is located at position 15q26.1. Patients with D2HGA have similar clinical features regardless of being type I or type II.

A third rare disorder involving 2HG has been described in 16 patients with severe neonatal-onset encephalopathy, in whom there were elevated levels of both L2HG and D2HG (D2L2AD OMIM # 615182) and of α-KG. Most of these infants with D2L2AD have died in the first five years of life. D2L2AD is caused by mutations in the SLC25A1 gene at chromosomal location 22q11.21. D2HGA1, L2HGA, and D2L2AD are inherited in an autosomal recessive man-ner, while D2HGA2 inheritance is autosomal dominant (Fig. 4.1).

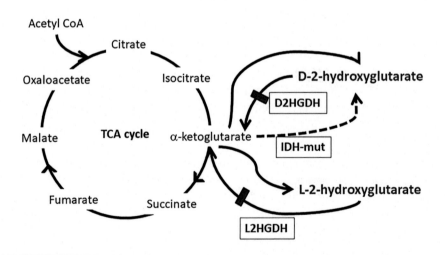

Figure 4.1 *Pathways involving 2-hydroxyglutaric acid.*

3 Clinical presentation

L2HGA is clinically characterized by childhood-onset neurodevelop-mental delay and subsequent variably progressive neurodegeneration with pyramidal and extrapyramidal findings, seizures, and ataxia. The ultimate developmental outcome is poor, with average intelligence quotients in the 40–50 range. Typical magnetic resonance imaging (MRI) findings include cortical atrophy, subcortical leukoencephalopathy, and high signal intensity in dentate nucleus and putamen. D2HGA may present with a severe phenotype of neonatal- or early infantile-onset epileptic encepha-lopathy and severe developmental delay, along with cardiomyopathy and facial dysmorphism. MRI findings include delayed cerebral maturation, ventricular white matter abnormalities, and subependymal cyst formation. D2HGA may also be asymptomatic or present with a milder phenotype of variable hypotonia and developmental delay along with milder MRI changes.

4 Diagnostic compounds
4.1 Urine organic acid profile

Large excretion of 2HG is the prominent finding along with TCA cycle intermediates. The gas chromatography-mass spectrometry method does not distinguish between the D- and L- stereoisomers of 2HG (Fig. 4.2 and Fig. 4.3).

4.2 Acylcarnitine profile

Non-diagnostic for 2HG.

4.3 Amino acids

Plasma, CSF, and urine lysine is elevated in patients with L2HGA.

4.4 Example chromatograph

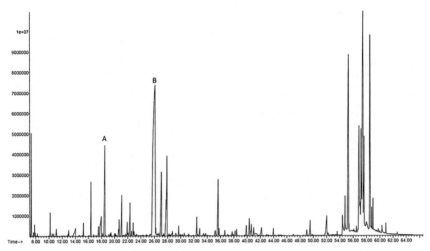

Figure 4.2 *D2HGA*. (A) Internal standard and (B) 2-hydroxyglutaric acid.

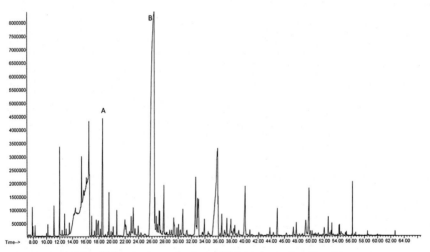

Figure 4.3 *L2HGA*. (A) Internal standard and (B) 2-hydroxyglutaric acid.

4.5 Other important diagnostic/monitoring compounds

None applicable.

5 Newborn screening

L2HGA, D2HGA, D2L2AD are not included as core conditions on the RUSP, nor as secondary conditions for newborn screening.

6 Follow-up/confirmatory testing

2-hydroxyglutaric aciduria is established by the presence of 2-hydroxyglutarate excretion in urine on organic acid analysis. The determination of L2HGA, D2HGA, or D2L2AD can be made by assays that differentiate the two stereoisomers. Since the 3 disorders have some distinct clinical features, diagnosis can be made based on clinical presentation and biochemical distinction is not always necessary. Genetic testing is also available.

7 Interferences and assay or interpretation quirks

2HG can also be elevated in GA2 (see Chapter 3. GA2). Artifactual presence of 2HG in urine can be seen from α-ketoglutarate degradation, bacterial contamination, uremia or lithium treatment.

Further reading

[1] NIH Genetic and Rare Diseases Information Center. Ornithine translocase deficiency. https://rarediseases.info.nih.gov/diseases/10761/2-hydroxyglutaric-aciduria   accessed 10/31/2019.
[2] Genetics Home Reference, NIH US National Library of Medicine. https://ghr.nlm.nih. gov/condition/2-hydroxyglutaric-aciduria accessed 10/31/2019.
[3] Kerr DS, Bedoyan JK. Disorders of pyruvate metabolism and the tricarboxylic acid cycle. In: Sarafoglou K, Hoffman GF, Roth KS, editors. Pediatric endocrinology and inborn errors of metabolism. 2nd ed. New York: McGraw-Hill Co; 2017. p. 105–24.
[4] Steenweg ME, Jakobs C, Errami A, van Dooren SJ, Adeva Bartolome MT, Aerssens P, Augoustides-Savvapoulou P, Baric I, Baumann M, Bonafe L, et al. An overview of L-2-hydroxyglutarate dehydrogenase gene (L2HGDH) variants: a genotype-phenotype study. Hum Mutat 2010;31:380–90.
[5] Struys EA. D-2-Hydroxyglutaric aciduria: unravelling the biochemical pathway and the genetic defect. J Inherit Metab Dis 2006;29:21–9.

CHAPTER 5

Disorder: Isovaleric aciduria

1 Synonyms

Isovaleric acidemia, IVA, isovaleryl-CoA dehydrogenase deficiency, isovaleric acid-CoA dehydrogenase deficiency, IVD deficiency.

2 Brief synopsis

2.1 Incidence

~1:80,000; 1:62,500 newborns in Germany; 1:250,000 in the United States.

2.2 Etiology

Isovaleric acidemia (IVA, OMIM #243500) is caused by a deficiency of the enzyme isovaleryl-CoA dehydrogenase (IVDH, EC 1.3.99.10). IVDH is a mitochondrial enzyme that catalyzes the oxidation of isovaleryl-CoA to 3-methylcrotonyl-CoA, a step in the catabolic pathway of the ketogenic branched-chain amino acid leucine (Fig. 5.1). IVDH is a 172 kDa enzyme which is composed of four identical peptides. Four mature 43 kDa peptides form a homo-tetramer with the dehydrogenase activity. Mutations result in reduced or absent enzyme activity and subsequent inability to properly catabolize leucine, resulting in a buildup of isovaleric acid and related compounds in tissues and body fluids. These excess compounds damage the brain and nervous system.

The gene that encodes IVDH is the *IVD* gene and it resides on chromosome 15q14-q15. At least 25 mutations have been identified in this gene in people with IVA, and it is inherited in an autosomal recessive fashion. A genotype/phenotype correlation is not seen, with both severe and mild presentations occurring within the same family and with the same mutation. This suggests that the environment and epigenetic factors play a role in this disorder. Through newborn screening programs a common mutation (c.932 c > t, p.A282V) has been identified that is associated with a very mild or asymptomatic clinical course.

A Quick Guide to Metabolic Disease Testing Interpretation
http://dx.doi.org/10.1016/B978-0-12-816926-1.00005-5

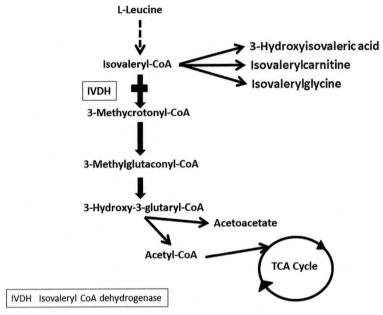

IVDH Isovaleryl CoA dehydrogenase

Figure 5.1 *Pathway involving isovaleryl-CoA dehydrogenase.*

3 Clinical presentation

Patients with IVA can be classified into two broad categories: those that present in early neonatal life with a sudden onset severe illness and those that have chronic intermittent disease and present later in infancy. The former presentation typically begins within a few days of birth with increasing lethargy and difficulty in feeding leading to dehydration and weight loss. An odor of sweaty feet may be found. The distinctive odor represents accumulation of isovaleric acid and related compounds in the body. Laboratory investigation typically shows metabolic acidosis with mild lactic acidemia and ketosis. There may be pancytopenia, hypocalcemia, and hyperammonemia. Many neonates do not survive the acute illness and die of acidosis, cerebral edema, infections, and/or bleeding. Those who do survive the initial acute phase, go on to follow a chronic intermittent course. Patients with the chronic intermittent form of the disease typically have had their first episode of acute illness by the time they are one year old. The acute episodes usually follow a minor stress like an upper respiratory infection and sometimes increased intake of protein-rich foods. The usual symptoms include lethargy, vomiting, and sweaty feet odor, with laboratory evidence of acidosis and pancytopenia. Some of these patients may have hyperglycemia, which along with ketosis can lead to a misdiagnosed diabetic ketoacidosis. The acute symptoms resolve with protein-restricted diet, and these children may learn to avoid protein-rich foods. This dietary strategy and fewer numbers of infections lead to fewer acute episodes, as the child grows older. Some children may have developmental delay and mental retardation, probably as a result of untreated acute events. Treatment involves a low-protein diet with addition of glycine and/or carnitine to ensure sufficient supplies for isovaleric acid conjugation and elimination from the body. Prognosis is significantly improved by rapid diagnosis and treatment to prevent the metabolic decompensations.

4 Diagnostic compounds

4.1 Urine organic acid profile

3-hydroxyisovaleric acid and isovalerylglycine excretions are elevated on urine organic acid analysis. Minor compounds that may sometimes be detected include 4-hydroxyvaleric acid, methylsuccinic acid, methylfumaric acid, 3-hydroxyisoheptanoic acid, isovalerylglutamic acid, isovalerylglucuronide, isovalerylalanine, and isovalerylsarcosine (Fig. 5.2).

4.2 Acylcarnitine profile

Isovalerylcarnitine (C5-carnitine) is usually markedly elevated. Free carnitine is often low.

4.3 Amino acids

Generally non-diagnostic for IVA.

4.4 Example chromatograph

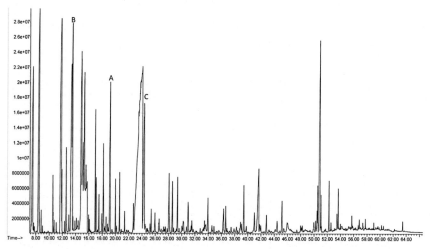

Figure 5.2 *Isovaleric aciduria.* (A) Internal standard, (B) 3-hydroxyisovaleric acid, (C) isovalerylglycine.

4.5 Other important diagnostic/monitoring compounds

None applicable.

5 Newborn screening

Isovaleric acidemia is part of the RUSP of newborn screening programs in the United States. Elevated concentrations of isovalerylcarnitine are detected in blood spots by tandem MS to screen for this disorder.

6 Follow-up/confirmatory testing

Diagnosis of IVA can be established by organic acid analysis for the elevated excretion of 3-hydroxyisovaleric acid and isovalerylglycine, in combination with acylcarnitine analysis for elevated concentrations of isovalerylcarnitine. Additional confirmation may be requested when asymptomatic newborns are picked up on newborn screening. Genetic testing for mutations in the *IVD* gene is also available.

7 Interferences and assay or interpretation quirks

3-hydroxisovaleric acid can be present due to ketosis. Additionally, valproate therapy can cause excretion of 3-hydroxisovaleric acid as well as isovalerylglycine.

Further reading

[1] NIH Genetic and Rare Diseases Information Center. Isovaleric Acidemia. https://rarediseases.info.nih.gov/diseases/465/isovaleric-acidemia accessed 11/1/2019.
[2] Genetics Home Reference, NIH US National Library of Medicine. https://ghr.nlm.nih.gov/condition/isovaleric-acidemia# accessed 11/1/2019.
[3] Hoffmann GF, Burlina A, Barshop BA. Organic acidurias. In: Sarafoglou K, Hoffman GF, Roth KS, editors. Pediatric endocrinology and inborn errors of metabolism. 2nd ed. New York: McGraw-Hill Co; 2017. p. 209–50.

CHAPTER 6

Disorder: 2-Methylbutyrylglycinuria

1 Synonyms

2-MBG, short/branched chain acyl-CoA dehydrogenase deficiency, SBCAD deficiency, 2-methylbutyryl-CoA dehydrogenase deficiency, 2-MBCD deficiency

2 Brief synopsis

2.1 Incidence

Extremely rare worldwide, unknown;
    1:250-500 in Hmong populations of Southeast Asia.

2.2 Etiology

2-methylbutyrylglycinuria (2MBG) (OMIM #600301) is an organic acid disorder of the isoleucine degradation pathway that is caused by absent or deficient concentrations of the enzyme 2-methylbutyryl-CoA dehydrogenase. Conversion of 2-methylbutyryl-CoA to tiglyl-CoA does not occur resulting in the buildup and excretion of 2-methylbutyrylglycine and 2-methylbutyrylcarnitine (Fig. 6.1).

2MBG deficiency, also called short/branch chain acyl-CoA dehydrogenase deficiency, is caused by mutations in the *ACADSB* gene located at position 10q26.13. More than 10 mutations in *ACADSB* have been identified in people with 2MBG deficiency. The disorder is inherited in an autosomal-recessive manner.

3 Clinical presentation

2MBG shows a wide range of possible clinical presentations. Many 2MBG individuals are clinically asymptomatic showing no health problems related to the diagnosis. The disorder however may present a few days after birth with poor feeding, irritability, vomiting, and lack of energy. Symptoms can also worsen to include difficulty breathing and even seizures or coma.

A Quick Guide to Metabolic Disease Testing Interpretation
http://dx.doi.org/10.1016/B978-0-12-816926-1.00006-7

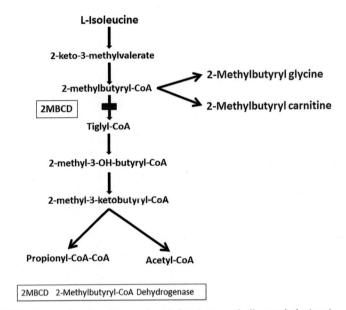

Figure 6.1 *Pathway showing the enzyme defect in 2-methylbutyryl glycinuria.*

Other symptoms that may be seen include poor growth, learning disabilities, muscle weakness, and vision impairment. It is unknown why some individuals show no symptoms and others may become very ill. If treatment is necessary, it usually involves a low protein diet and careful observation during illness to prevent metabolic decompensation.

4 Diagnostic compounds

4.1 Urine organic acid profile

2-Methylbutyrylglycine is present.

4.2 Acylcarnitine profile

C5–carnitine is elevated.

4.3 Amino acids

Non-diagnostic for 2MBG.

4.4 Example chromatograph

Fig. 6.2.

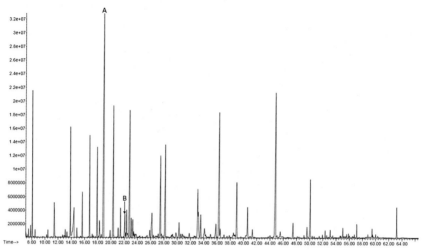

Figure 6.2 *2-methylbutyrylglycinuria.* (A) Internal standard, (B) 2-methylbutyrylglycine.

4.5 Other important diagnostic/monitoring compounds

None applicable.

5 Newborn screening

2MBG is not included as a core condition on the RUSP for newborn screening, but is included as a secondary condition, as it will be diagnosed in the workup for the differential diagnosis of an elevated C5-carnitine.

Tandem mass spectrometry (MS/MS) detection of elevated blood levels of the C5-carnitine is seen in both isovaleric acidemia and 2MBG.

6 Follow-up/confirmatory testing

Diagnosis of 2MBG can be established by the presence of 2-methylbutyryl glycine in the urine on organic acid analysis, especially when accompanied by an elevated C5-carnitine on acylcarnitine analysis. The diagnosis can be confirmed by molecular testing for mutations in the *ACADSB* gene.

7 Interferences and assay or interpretation quirks

Elevated C5-carnitine on acylcarnitine analysis is seen in both isovaleric acidemia and 2MBG, so acylcarnitine analysis alone is insufficient to differentiate between these two disorders. Additionally, elevated C5-carnitine may be seen when valproate or some antibiotics are being taken. Urine organic acid analysis is necessary to differentiate the causes of C5-carnitine elevations.

Further reading

[1] NIH Genetic and Rare Diseases Information Center. 3-Methylbutyryl-CoA dehydroge-nase deficiency. https://rarediseases.info.nih.gov/diseases/10322/short-branched-chain-acyl-coa-dehydrogenase-deficiency accessed 12/16/2019.
[2] Genetics Home Reference, NIH US National Library of Medicine. https://ghr.nlm.nih.gov/condition/short-branched-chain-acyl-coa-dehydrogenase-deficiency#   accessed 12/16/2019.
[3] Gibson KM, Burlingame TG, Hogema B, Jakobs C, Schutgens R BH, Millington D, Roe CR, Roe DS, Sweetman L, Steiner RD, Linck L, Pohowalla P, Sacks M, Kiss D, Rinaldo P, Vockley J. 2-Methylbutyryl-coenzyme A dehydrogenase deficiency: a new inborn error of L-isoleucine metabolism. Pediat Res 2000;47:830–3.
[4] Andresen BS, Christensen E, Corydon TJ, Bross P, Pilgaard B, Wanders RJA, Ruiter JPN, Simonsen H, Winter V, Knudsen I, Schroeder LD, Gregersen N, Skovby F. Isolated 2-methylbutyrylglycinuria caused by short/branched-chain acyl-CoA dehydrogenase deficiency: identification of a new enzyme defect, resolution of its molecular basis, and evidence for distinct acyl-CoA dehydrogenases in isoleucine and valine metabolism. Am J Hum Genet 2000;67:1095–103.

CHAPTER 7

Disorder: 3-Methylcrotonyl-CoA-carboxylase deficiency

1 Synonyms

3MCC, 3-methylcrotonylglycinuria.

2 Brief synopsis

2.1 Incidence

1:50,000

2.2 Etiology

3MCC deficiency (OMIM #210200, 210210) is reported to be the most frequently diagnosed organic aciduria in newborn screening programs. 3-Methylcrotonylglycinuria is a disorder of the catabolic pathways of the amino acid L-leucine. 3-Methylcrotonyl-CoA-carboxylase (EC6.4.1.4) activity is deficient, preventing the conversion of 3-methylcrotonyl-CoA to 3-methylglutaconyl-CoA. The lack of enzyme activity results in the accumulation of 3-methylcrotonylglycine, 3-hydroxyisovaleric acid, and 3-hydroxyisovalerylcarnitine (Fig. 7.1). This enzyme requires biotin for activity, so 3-methylcrotonylglycinuria can also be caused by primary deficiencies in the biotin pathways (see Chapter 33, Biotin). Defects in the biotin pathways will also affect the other biotin-dependent carboxylases—propionyl-CoA carboxylase, pyruvate carboxylase, and acetyl-CoA carboxylase.

Isolated 3MCC deficiency is caused by mutations in the *MCCC1* or *MCCC2* genes, both of which are responsible for encoding a part of the enzyme 3-methylcrotonyl-CoA carboxylase. *MCCC1* encodes for the alpha subunit and its location is 3q27.1. *MCCC2* encodes for the beta subunit and its location is 5q13.2. At least 30 mutations in *MCCC1* and 40 mutations in *MCCC2* have been identified in people with 3MCC. The disorder is inherited in an autosomal-recessive manner.

A Quick Guide to Metabolic Disease Testing Interpretation
http://dx.doi.org/10.1016/B978-0-12-816926-1.00007-9

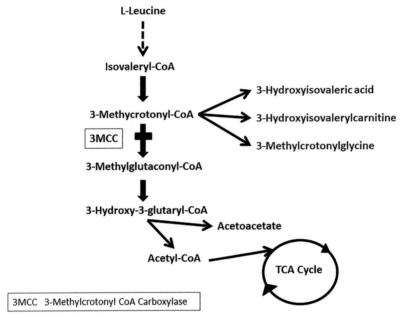

Figure 7.1 *Pathway showing the enzyme defect in 3-MCC deficiency.*

3 Clinical presentation

This disorder shows a wide range of possible clinical presentations from apparently clinically benign with no symptoms at all to a fulminant course that may result in death. The disorder has been picked up in women with no previous symptoms by means of detecting the associated biomarkers in their infant on newborn screen. A significant number of individuals with 3MCC never develop any symptoms. However, 3MCC deficiency may present in crises, usually triggered by infections between 6 months and 3 years of age. During crisis, patients may show vomiting, feeding difficulties, hypotonia, hyperreflexia, spasms, and seizures. Laboratory values will often show severe hypoglycemia, hyperammonemia, mild metabolic acidosis with moderate ketosis, and elevated transaminases during these episodes. Severe crises may result in death due to cerebral edema or cardiopulmonary arrest.

Treatment involves prevention of metabolic decompensation and appropriate treatment during acute decompensations, including providing increased energy through oral or IV glucose and fats.

4 Diagnostic compounds

4.1 Urine organic acid profile

3-Hydroxyisovaleric acid is always present and 3-methylcrotonylglycine will range between significantly elevated and barely detectable. 3-Hydroxyisovaleric acid is a very nonspecific metabolite that is often present even in normal urine. By itself it has no single interpretation; thus, looking for other metabolites and patterns of metabolites is necessary. If 3-hydroxypropionic acid, methylcitric acid, and/or lactic acid are also present, consider the possibility of multiple-carboxylase deficiency.

4.2 Acylcarnitine profile

C5-OH-carnitine is always elevated. This compound may be elevated in association with other acylcarnitines in other disorders (multiple carboxylase deficiency, β-ketothiolase deficiency, HMG CoA lyase deficiency, and 2-methy-3-hydroxybutyrl-CoA dehydrogenase deficiency). A urine organic acid profile is necessary to distinguish among these disorders.

4.3 Amino acids

Non-diagnostic for 3MCC deficiency.

4.4 Example chromatograph

Fig. 7.2 and Fig. 7.3.

4.5 Other important diagnostic/monitoring compounds

None applicable.

5 Newborn screening

3MCC is included on the RUSP for newborn screening.

Newborn screening for 3MCC is based on tandem mass spectrometry (MS/MS) detection of elevated blood levels of the C5-OH carnitine. Free and total carnitine are often low.

6 Follow-up/confirmatory testing

Diagnosis of 3MCC deficiency can be confirmed by demonstrating reduced enzyme activity in fibroblasts or leukocytes without response to biotin. Demonstration of disease-causing mutations by molecular testing is also available.

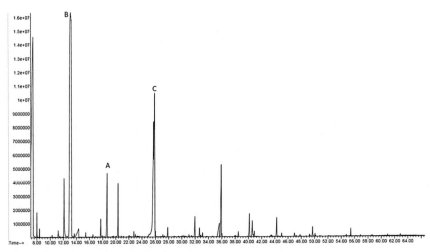

Figure 7.2 *3-methylcrotonylglycinuria* . (A) Internal standard, (B) 3-hydroxyisovaleric acid, and (C) 3-methylcrotonylglycine.

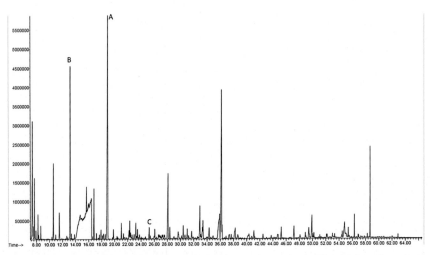

Figure 7.3 *3-methylcrotonylglycinuria* . (A) Internal standard, (B) 3-hydroxyisovaleric acid, and (C) 3-methylcrotonylglycine.

7 Interferences and assay or interpretation quirks

Elevated C5-OH-carnitine on acylcarnitine analysis and the presence of 3-methylcrotonylglycine on organic acid analysis may be present in the newborn if the mother has 3MCC deficiency. Additionally, infants with vegan mothers or biotin deficiency display these biochemical abnormalities that mimic 3-MCC. Therefore, repeat measurements with increasing concentrations of C5-OH carnitine are necessary to establish a 3MCC diagnosis in the neonate. Repeat measurements with decreasing concentrations suggest maternal 3MCC deficiency as the source of these biomarkers.

Further reading

[1] NIH Genetic and Rare Diseases Information Center. 3-Meyhlycrotonyl-CoA carboxylase deficiency. https://rarediseases.info.nih.gov/diseases/10954/3-methylcrotonyl-coa-carboxylase-deficiency accessed 11/21/2019.
[2] Genetics Home Reference, NIH US National Library of Medicine. https://ghr.nlm.nih.gov/condition/3-methylcrotonyl-coa-carboxylase-deficiency accessed 11/21/2019.
[3] Hoffmann GF, Burlina A, Barshop BA. Organic acidurias. In: Sarafoglou K, Hoffman GF, Roth KS, editors. Pediatric endocrinology and inborn errors of metabolism. 2nd ed. New York: McGraw-Hill Co; 2017. p. 209–50.

CHAPTER 8

Disorder: 3-Methyglutaconic aciduria

1 Synonyms

3MGA, 3-methylglutaconyl-CoA hydratase deficiency (type 1), Barth syndrome (type 2); Costeff optic atrophy (type3), 3MGA type 4, 3MGA type 5

2 Brief synopsis

2.1 Incidence

<1:200,000

2.2 Etiology

3-Methylglutaconic aciduria (3MGA) is a biochemical finding that describes a heterogeneous group of at least nine disorders, including 3-methylglutaconyl-CoA hydratase deficiency (3MGA type I, 3MGA1) (OMIM #250950), Barth syndrome (3MGA type II, 3MGA2) (OMIM #302060), Costeff optic atrophy (3MGA type III, 3MGA3) (OMIM #258501), 3MGA, type IV (3MGA4) (OMIM #250951) and 3MGA, type V (3MGA5) (OMIM #610198). These disorders all show elevated excretion of 3-methylglutaconic acid and 3-methylglutaric acid in the urine. Additionally, several mitochondrial and respiratory chain disorders also show elevated excretion of 3-methylglutaconic acid, such as Pearson's syndrome, Leigh syndrome and ATP synthetase deficiency. Type I is due to a deficiency in 3-methylgluaconyl-CoA hydratase, which is required for the conversion of 3-methylglutaconyl-CoA to 3-hydroxy-3-methylglutaryl-CoA (important in leucine catabolism) (Fig. 8.1). Types II and V are caused by defects in different inner mitochondrial membrane proteins, and the mechanism of 3-methylglutaconic and 3-methylglutaric acid excretion is unknown. Type III is caused by defects in a protein with unknown function found in the mitochondria. Type IV is a heterogeneous group of individuals with no defined cause who excrete elevated amounts of 3-methylglutaconic acid and 3-methylglutaric acid into the urine continuously or intermittently.

A Quick Guide to Metabolic Disease Testing Interpretation
http://dx.doi.org/10.1016/B978-0-12-816926-1.00008-0

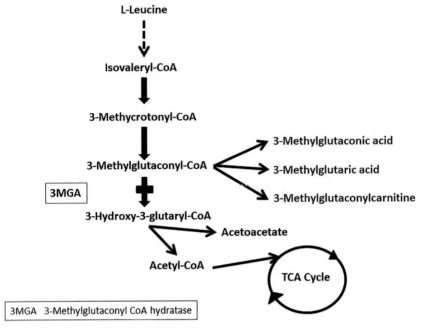

Figure 8.1 *Pathway showing the enzyme defect in 3MGA1.*

There are at least 11 known mutations in the *AUH* gene at location 9q22.31 that result in 3MGA type I. Type II has more than 160 known mutations in the *TAZ* gene located at Xq28, and is the only 3MGA disorder with *X*-linked inheritance. The other MGA disorders show autosomal-recessive inheritance. Type III is caused by at least 5 mutations in *OPA3* gene located at 19q12.32 and Type V is caused by mutations in the *DNAJC19* gene at chromosomal location 3q26.33.

3 Clinical presentation

This group of disorders shows a wide range of possible clinical presentations due to the heterogeneity of the underlying causes of excess excretion of 3MGA into the urine. 3MGA type I presents in infancy to early childhood with developmental delay, muscle cramping and spastic quadriparesis. Optic atrophy may occur, and these children may go on to develop leukoencephalopathy. In cases that present in adulthood, the leukoencephalopathy causes progressive speech and movement difficulties, spasticity, optic atrophy and dementia. Asymptomatic individuals with type I have been reported. Type II generally presents at birth or shortly thereafter with dilated cardiomyopathy, skeletal myopathy, fatigue and neutropenia. Males with type II have a shortened life expectancy. The defining feature of type III is optic atrophy beginning in infancy. Spasticity, ataxia and choreiform movements develop in later childhood, and spasticity may be progressive. Intelligence may be normal or show up to moderate impairment. 3MGA5 is also called dilated cardiomyopathy with ataxia syndrome, and the major characteristics are described by this name. Growth is slow in these individuals and males may have genital abnormalities. Many do not survive childhood due to heart failure. Type IV is the most heterogeneous group as it encompasses all patients with elevated excretion of 3MGA who do not fall into the clearly defined causes. In general individuals with type IV show some combination of neurological abnormalities, developmental abnormalities, myopathy and cardiomyopathy. Treatment for 3MGA involves treating the symptoms.

4 Diagnostic compounds

4.1 Urine organic acid profile

3-Methylglutaconic acid and 3-methylglutaric acid are present in elevated concentrations in the urine for all forms of 3-methylglutaconic aciduria. Type I often also shows elevated amounts of 3-hydroxyisovaleric acid. Type II also often shows elevated amounts of 2-ethylhydracrylic acid.

4.2 Acylcarnitine profile

3-Methylglutaconylcarnitine (C6DC-carnitine) may be elevated in cases of type I.

4.3 Amino acids

Non-diagnostic for 3MGA.

4.4 Example chromatograph

Fig. 8.2 and Fig. 8.3.

4.5 Other important diagnostic/monitoring compounds

None applicable.

5 Newborn screening

3MGA is not included as a core condition on the RUSP for newborn screening, but it is included as a secondary condition that will be diagnosed in the workup of the core condition 3-hydroxy-3-methylglutaric aciduria

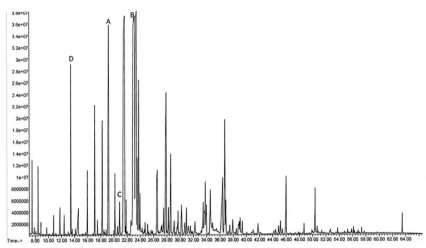

Figure 8.2 *3-Methylglutaconic aciduria type I* . (A) Internal standard, (B) 3-methylgluta-conic acid, (C) 3-methylglutaric acid, and (D) 3-hydroxyisovaleric acid.

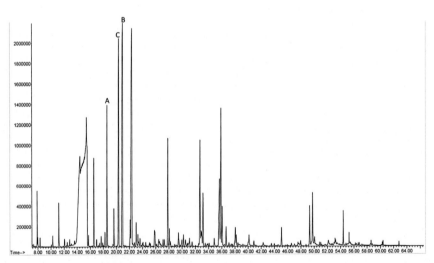

Figure 8.3 *3-Methylglutaconic aciduria type IV* . (A) Internal standard, (B) 3-methylglu-taconic acid, (C) 3-methylglutaric acid.

(HMG). Both of these conditions show an elevated C6DC-carnitine in dried blood spots by tandem mass spectrometry (MS/MS).

6 Follow-up/confirmatory testing

Diagnosis of 3MGA is established by demonstrating continuously elevated excretion of 3-methylglutaconic acid and 3-methylglutaric acid in urine. Molecular testing of the specific genes for mutations can be done to confirm diagnosis, guided by clinical symptoms for the various types of 3MGA.

7 Interferences and assay or interpretation quirks

3-methylglutanoic acid can be non-specifically elevated in a urine organic acid profile due to many non-IEM causes including uremia, acquired HMG-CoA lysase deficiency and pregnancy.

Further reading

[1] NIH Genetic and Rare Diseases Information Center. 3-Methylglutaconyl-CoA hydratase deficiency. https://rarediseases.info.nih.gov/diseases/10321/3-alpha-methylglutaconic-aciduria-type-i accessed 12/16/2019.

[2] Genetics Home Reference, NIH US National Library of Medicine. https://ghr.nlm.nih.gov/condition/3-methylglutaconyl-coa-hydratase-deficiency#   accessed 12/16/2019.

[3] NIH Genetic and Rare Diseases Information Center. Barth Syndrome. https://rarediseases.info.nih.gov/diseases/5890/barth-syndrome accessed 12/16/2019.

[4] Genetics Home Reference, NIH US National Library of Medicine. https://ghr.nlm.nih.gov/condition/barth-syndrome# accessed 12/16/2019.

[5] GeneReviews, NIH, Barth Syndrome. https://www.ncbi.nlm.nih.gov/books/NBK247162/accessed 12/6/2019.

[6] NIH Genetic and Rare Diseases Information Center. Costeff Syndrome. https://rarediseases.info.nih.gov/diseases/5663/costeff-syndrome accessed 12/16/2019.

[7] Genetics Home Reference, NIH US National Library of Medicine. https://ghr.nlm.nih.gov/condition/costeff-syndrome accessed 12/16/2019.

[8] GeneReviews, NIH, OPA3-Related 3-methylgutaconic aciduria. https://www.ncbi.nlm.nih.gov/books/NBK1473/accessed 12/6/2019.

[9] Genetics Home Reference, NIH US National Library of Medicine. https://ghr.nlm.nih.gov/condition/dilated-cardiomyopathy-with-ataxia-syndrome accessed 12/16/2019.

[10] Hoffmann GF, Burlina A, Barshop BA. Organic acidurias. In: Sarafoglou K, Hoffman GF, Roth KS, editors. Pediatric endocrinology and inborn errors of metabolism. 2nd ed. New York: McGraw-Hill Co; 2017. p. 209–50.

CHAPTER 9

Disorder: Methylmalonic aciduria

1 Synonyms

MMA, methylmalonic acidemia

2 Brief synopsis

2.1 Incidence

1:100,000 for mutase deficiency.

2.2 Etiology

The accumulation of methylmalonic acid in body fluids is the common biochemical feature of a number of inborn errors of metabolism, including methylmalonyl-CoA mutase deficiency (OMIM #251000) and a large number of cobalamin metabolic defects: CblA – OMIM #251100, CblB – OMIM #251110, CblC – OMIM #277400, CblD – OMIM #277410 and the much more rare disorders CblE, CblF, CblG, ClbJ, CblX.

A deficiency in methylmalonyl-CoA mutase (EC5.4.99.2) causes the inability to convert methylmalonyl-CoA to succinyl-CoA, resulting in a backup of the metabolites of methylmalonic acid and propionic acid (Fig. 9.1). Defects in intracellular cobalamin metabolism are divided into complementation classes. CblA and CblB defects are involved in adenosylcobalamin metabolism, affect the mutase enzyme, and result in similar patterns of biochemical metabolites seen with mutase deficiency. Like the mutase deficiency, they are picked up in the expanded newborn screen utilizing tandem mass spectrometry. CblC and CblD defects appear to affect methylcobalmin and adenosylcobalamin metabolism, resulting in impaired activity of the mutase and the conversion of homocystine to methionine by methionine synthase. Methylmalonic acid and homocystine will be elevated (Fig. 9.2). The CblC defect is the most common inborn error of cobalamin metabolism. A defect in CblF is a very rare, and it causes a disorder that involves trapping of the cobalamin in the lysosomes after transcobalamin degradation, making it unavailable as adenoslycobalamin or methylcobalmin. CblE and Cb1G are caused by mutations in the methionine synthase reductase and methionine synthase genes, respectively.

A Quick Guide to Metabolic Disease Testing Interpretation
http://dx.doi.org/10.1016/B978-0-12-816926-1.00009-2

51

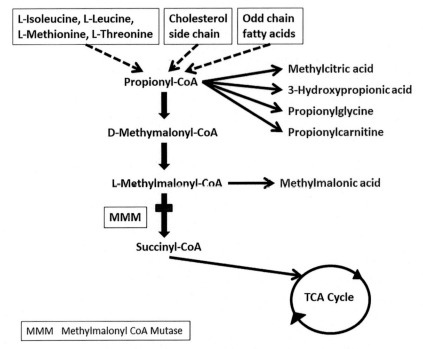

Figure 9.1 *Pathway showing the methylmalonyl-CoA mutase defect in MMA.*

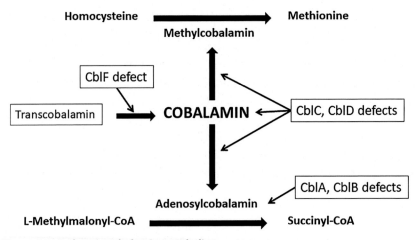

Figure 9.2 *Defects in cobalamin metabolism.*

Methylmalonyl–CoA mutase deficiency is caused by mutations in the *MUT* gene and is inherited in an autosomal recessive manner. The gene's location is 6p12.3, and more than 200 mutations have been identified in individuals with MMA. The most severe form of MMA is the result of a Mut^0 mutation resulting in no functional mutase being produced.

Mutations in the *MMAA* and *MMAB* genes, at locations 4q31.21 and 12q24.11 respectively, result in CblA and CblB deficiencies by causing adenosylcobalamin to be defective. *MMACHC* and *MMADHC* gene mutations result in CblC and CblD deficiencies by interfering with proper processing and transport of both adenosylcobalamin and methylcobalamin. These genes are located at 1p34.1 and 2q23.2, respectively.

3 Clinical presentation

Methylmalonyl-CoA mutase deficiency usually presents within the first months of life with severe metabolic acidosis, ketosis, and lactic acidemia. Clinical manifestations may be variable; however, almost all present with failure to thrive, poor growth, feeding problems, and some developmental delay. Metabolic crisis is often accompanied by vomiting, lethargy, seizures, and hepatomegaly. CblA, CblB, CblG, and CblE defects may present with very similar clinical findings to the mutase deficiencies. CblC, CblD, and CblF may present as acute neurological deterioration and multisystem disease. Encephalopathy and cerebral atrophy are also seen in diseases of cobalamin metabolism, as is megaloblastic anemia, and bone marrow failure is seen with CblA and CblB defects.

Treatment of MMA caused by cobalamin defects usually includes vitamin B12 supplementation in the form of hydroxycobalamin. Patients are more or less responsive depending on the specific defect. Mutase deficiency MMA requires a protein-restricted diet, especially low in the branched chain amino acids which are metabolized to propionyl- CoA. All types of MMA must be treated rapidly and aggressively when metabolic decompensation occurs. Prognosis is better for cobalamin-responsive forms of MMA.

4 Diagnostic compounds

4.1 Urine organic acid profile

Elevated concentrations of methylmalonic acid and detectable or elevated methyl citrate are always present. Depending on the severity of the mutase deficiency, 3-hydroxypropionic acid, propionylglycine and tiglylglycine may also be present.

4.2 Acylcarnitine profile

Propionyl- or C3-carnitine is elevated in both MMA and propionic acidemia. Free carnitine is often low.

4.3 Amino acids

In MMA caused by cobalamin defects, homocysteine will also be elevated and methionine will be low.

4.4 Example chromatographs

Fig. 9.3 and Fig. 9.4.

4.5 Other important diagnostic/monitoring compounds

Ammonia and lactic acid should be monitored as they will often be elevated during crises.

5 Newborn screening

MMA is included on the RUSP for newborn screening for both the mutase deficiencies and the cobalamin defects. Newborn screening for MMA is based on tandem mass spectrometry (MS/MS) detection of elevated blood levels of the C3 carnitine and an elevated C3/C0 ratio. CblC and CblD defects are often not picked up in the expanded newborn screen because

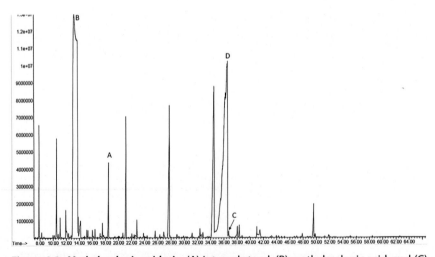

Figure 9.3 *Methylmalonic aciduria.* (A) Internal stand, (B) methylmalonic acid, and (C) methylcitrate at base of a large peak of (D) hippurate.

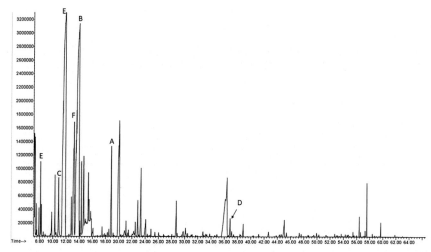

Figure 9.4 *Methylmalonic aciduria.* (A) Internal stand, (B) methylmalonic acid, (C) 3-hydroxypropionic acid, (D) methylcitrate, and the ketones 3-hydroxybutyric acid, (E) and acetoacetate (F).

they tend not to elevate propionylcarnitine above the newborn screening cutoff within the first 48 hours of life.

6 Follow-up/confirmatory testing

Diagnosis of MMA can be confirmed by demonstrating continuously elevated C3-carnitine on acylcarnitine analysis, along with an organic acid analysis that shows increased excretion of methylmalonic acid and the presence of methylcitrate. Elevated homocysteine concentrations along with these finding are consistent with cobalamin defects rather than mutase defects. Molecular testing is available to help distinguish among CblC, CblD, and CblF types.

7 Interferences and assay or interpretation quirks

Propionic acidemia (PA) shows the same acylcarnitine profile as MMA, with an elevated C3-carnitine. A urine organic acid analysis is necessary to help distinguish among PA and MMA. In MMA, methylmalonic acid and methylcitrate are present. In PA, larger amounts of 3-hydroxypropionic acid and propionylglycine are present, along with methylcitrate and often tiglylglycine, and methylmalonic acid is usually small or absent.

Non-IEM causes of methylmalonic aciduria include vitamin B12 deficiency, pernicious anemia, bacterial metabolism, gastroenteritis in young infants, short bowel syndrome and malnutrition. 3-hydroxypropionic excretion can be caused by lactic acidosis, short bowel syndrome and bacterial metabolism.

Further reading

[1] NIH Genetic and Rare Diseases Information Center. Methylmalonic Acidemia. https://rarediseases.info.nih.gov/diseases/7033/methylmalonic-acidemia accessed 11/22/2019.

[2] Genetics Home Reference, NIH US National Library of Medicine. https://ghr.nlm.nih.gov/condition/methylmalonic-acidemia accessed 11/22/2019.

[3] GeneReviews, NIH, Isolated Methylmalonic Acidemia. https://www.ncbi.nlm.nih.gov/books/NBK1231//accessed 11/22/2019.

[4] GeneReviews, NIH, Disorders of Intracellular Cobalamin Metabolism. https://www.ncbi.nlm.nih.gov/books/NBK1328/accessed 11/22/2019.

[5] Hoffmann GF, Burlina A, Barshop BA. Organic acidurias. In: Sarafoglou K, Hoffman GF, Roth KS, editors. Pediatric endocrinology and inborn errors of metabolism. 2nd ed. New York: McGraw-Hill Co; 2017. p. 209–50.

[6] Watkins D, Morel CF, Rosenblatt DS. Inborn errors of folate and cobalamin transport and metabolism. In: Sarafoglou K, Hoffman GF, Roth KS, editors. Pediatric endocrinology and inborn errors of metabolism. 2nd ed. New York: McGraw-Hill Co; 2017. p. 287–307.

CHAPTER 10

Disorder: Propionic acidemia

1 Synonyms

PA, Propionyl-CoA carboxylase deficiency, PCC deficiency, ketotic hyper-glycinemia

2 Brief synopsis

2.1 Incidence

1:100,000–200,000.

2.2 Etiology

Propionic acidemia (PA; OMIM #232000) is caused by a deficiency in propionyl-CoA carboxylase (EC6.4.1.3), the enzyme that coverts propi-onyl-CoA to D-methylmalonyl-CoA. Propionyl-CoA is an intermediate byproduct of the catabolism of many compounds in the body, including the branched-chain amino acids leucine, isoleucine, and valine; the amino acids threonine and methionine; the odd-chain fatty acids; and the side chain of cholesterol. Propionyl-CoA accumulates when there are defects in propi-onyl-CoA carboxylase, resulting in accumulations of 3-hydroxypropionic acid, methyl citrate, propionylglycine, propionylcarnitine, and propionic acid (Fig. 10.1). Propionic acid is usually not measurable, perhaps because of being rapidly excreted as propionylglycine and 3-hydroxypropionic acid, and because of propionyl-CoA being shunted into other pathways (i.e., entering the Krebs cycle to form methyl citrate). There is some thought that these accumulated compounds are toxic and/or inhibit other metabolic pathways, resulting in hyperglycinemia, hyperammonemia, ketosis, and lac-tic acidosis.

Propionyl-CoA carboxylase is composed of twelve subunits, 6 alpha subunits and 6 beta subunits, which are encoded by the *PCCA* and *PCCB* genes, located at positions 13q32.3 and 3q22.3, respectively. Over 120 mu-tations in *PCCA* and 100 mutations in *PCCB* have been identified in in-dividuals with PA, and the disorder has an autosomal-recessive mode of inheritance. Like other biotin-dependent carboxylases, defects in the biotin

A Quick Guide to Metabolic Disease Testing Interpretation
http://dx.doi.org/10.1016/B978-0-12-816926-1.00010-9

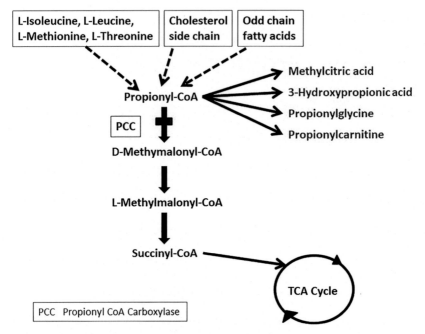

Figure 10.1 *Pathway involving propionyl-CoA carboxylase.*

pathways, including biotinidase and holocarboxylase synthase, will also affect propionyl–CoA carboxylase (see Chapter 33, Biotin).

3 Clinical presentation

Propionic acidemia often presents in the neonatal period with massive metabolic decompensation. Presentation in these cases is usually rapid, extreme, and progressive, and without correct diagnosis and treatment, it is often fatal. The presenting symptoms of overwhelming illness are nonspecific and suggest sepsis, delaying the diagnosis unless a metabolic workup accompanies the sepsis workup. Severe hyperammonemia is usually present, along with metabolic acidosis with an elevated anion gap, lactic acidosis, hyperglycinemia, elevated alanine, and ketosis. Initially the disorder was included in the spectrum of diseases called ketotic hyperglycinemia. Past the neonatal stage, metabolic decompensations often occur frequently, with similar presentation to those of the neonatal stage along with vomiting, hypotonia, seizures, and hepatomegaly. Growth may be impaired in the long term. Neurological symptoms are common, with most patients exhibiting developmental delay, seizures, and mental retardation.

Like other organic acidemias, rapid emergency treatment of decompensations is the best predictor of outcome. Long-term treatment involves a diet with low levels of the branched chain amino acids and methionine and threonine. Decompensations are accompanied by massive ketoacidosis and hyperammonemia, both of which must be treated aggressively.

4 Diagnostic compounds

4.1 Urine organic acid profile

Elevated concentrations of 3-hydroxypropionic acid, methylcitrate, propionylglycine, and tiglylglycine are the most common finding in propionic acidemia. Lactic acid is often elevated (Fig. 10.2).

4.2 Acylcarnitine profile

Propionylcarnitine (C3-carnitine) is usually massively elevated and used as a marker for monitoring treatment in patients with PA

4.3 Amino acids

Generally non-diagnostic for PA although glycine may be elevated.

4.4 Example chromatograph

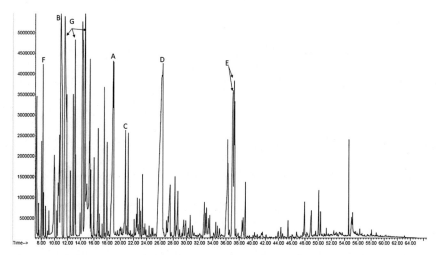

Figure 10.2 *Propionic acidemia.* (A) Internal stand, (B) 3-hydroxypropionic acid, (C) propionylglycine, (D) tiglylglycine (E) two peaks of methylcitrate, (F) lactic acid, and (G) ketones (3-hydroxybutyric acid, acetoacetate).

4.5 Other important diagnostic/monitoring compounds

Ammonia and ketones should be monitored since these individuals are prone to significant hyperammonemia and ketoacidosis.

5 Newborn screening

Propionic acidemia is part of the RUSP of newborn screening programs in the United States. Elevated concentrations of propionylcarnitine are detected in blood spots by tandem MS to screen for this disorder and the C3/C2 ratio is generally elevated.

6 Follow-up/confirmatory testing

Diagnosis of PA can be established by organic acid analysis for the elevated excretion of 3-hydroxypropionic acid, methylcitrate and propionylglycine, in combination with acylcarnitine analysis for elevated concentrations of propionylcarnitine. Enzyme analysis testing is available on fibroblasts or lymphocytes, and molecular genetic analysis for mutations in the *PCCA* and *PCCB* genes are also available.

7 Interferences and assay or interpretation quirks

Defects of the biotin pathways result in multiple carboxylase deficiency and will also show a urine organic acid profile containing elevated excretion of 3-hydroxypropionic acid, methylcitrate and propionyl glycine, however 3-hydroxyisovaleric acid and 3-methylcrotonylglycine will also be present. (See Chapter 33, Biotin). Non–IEM causes of 3-hydroxypropionic excretion include lactic acidosis, short bowel syndrome and bacterial metabolism. Non–IEM causes of 3-hydroxyisovaleric acid include ketosis and valproate treatment. Reye syndrome can also result in excretion of 3-hydroisovaleric acid and 3-methylcrotonylglycine.

Further reading

[1] NIH Genetic and Rare Diseases Information Center. Propionic Acidemia. https://rare-diseases.info.nih.gov/diseases/467/propionic-acidemia accessed 11/27/2019.
[2] Genetics Home Reference, NIH US National Library of Medicine. https://ghr.nlm.nih.gov/condition/propionic-acidemia# accessed 11/27/2019.
[3] GeneReviews, NIH, Propionic Acidemia. https://www.ncbi.nlm.nih.gov/books/NBK92946/accessed 11/27/2019.
[4] Hoffmann GF, Burlina A, Barshop BA. Organic acidurias. In: Sarafoglou K, Hoffmann GF, Roth KS, editors. Pediatric endocrinology and inborn errors of metabolism. 2nd ed. New York: McGraw-Hill Co; 2017. p. 209–50.

CHAPTER 11

Disorder: Succinic semialdehyde dehydrogenase deficiency

1 Synonyms

SSADH, 4-hydroxybutyric aciduria, gamma-hydroxybutyric acidemia

2 Brief synopsis

2.1 Incidence

<1:200,000

2.2 Etiology

Succinic semialdehyde dehydrogenase (SSADH) deficiency (OMIM #271980) is a neurotransmitter disorder caused by a deficiency in the enzyme SSADH (EC1.2.1.24), which catalyzes the conversion of γ-amino butyric acid (GABA) to succinic acid. Deficiency in this enzyme results in the accumulation of GABA and 4-hydroxybutyric acid, also called γ-hydroxybutyrate (GHB) (Fig. 11.1). Both of these compounds are neurotransmitters, and SSADH deficiency is one of the few inherited disorders in which there is an accumulation of compounds that are known to be neuropharmacologically active. It is unclear how accumulation of these neurotransmitters leads to the clinical symptoms seen with this disorder. The defining feature of SSADH deficiency is the accumulation of 4-hydroxybutyric acid in physiological fluids, but no metabolic acidosis is present.

SSADH deficiency is caused by mutations in the *SSADH* gene, now called the *ALDH5A1* gene at location 6p22.3. At least 35 mutations have been shown to cause SSADH deficiency. The disorder is inherited in an autosomal recessive manner.

3 Clinical presentation

Presentations seen with SSADH deficiency are heterogeneous. The disorder tends to present in late infancy or early childhood with delayed language development, especially expressive language, mild to moderate

A Quick Guide to Metabolic Disease Testing Interpretation
http://dx.doi.org/10.1016/B978-0-12-816926-1.00011-0

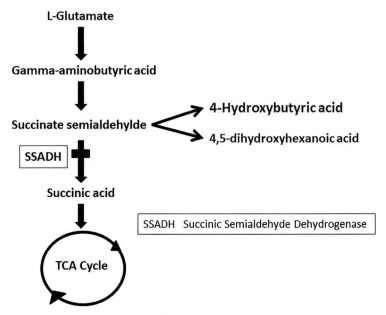

L-Glutamate

Gamma-aminobutyric acid

Succinate semialdehylde

4-Hydroxybutyric acid

4,5-dihydroxyhexanoic acid

SSADH

Succinic acid

| SSADH | Succinic Semialdehyde Dehydrogenase |

TCA Cycle

Figure 11.1 *Pathway involving succinic semialdehyde dehydrogenase.*

psychomotor retardation, ataxia, hypotonia, behavioral disturbances, and occasionally seizures. Encephalopathy is slowly progressive. Hypotonia and occasionally ataxia seen in many patients appears to resolve somewhat with age. A diagnosis of autism spectrum disorder may be made. Behavioral disturbances include aggression, sleep disturbances, self-injurious behavior, and occasionally hallucinations. Involvement outside the CNS has not been seen.

Treatment and management is usually directed at the seizures and behavioral disturbances. Prognosis is guarded, as the encephalopathy is slowly progressive, but appears mostly static. Life expectancy is not reduced.

4 Diagnostic compounds

4.1 Urine organic acid profile

Elevated concentrations of 4–hydroxybutyric acid are always present, sometimes with increased excretion of 4,5–dihydroxyhexanoic acid.

4.2 Acylcarnitine profile

Non-diagnostic for SSADH.

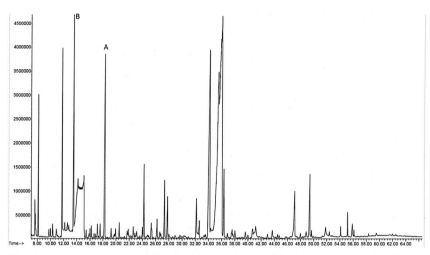

Figure 11.2 *Succinic semialdehyde dehydrogenase deficiency.* (A) Internal standard and (B) 4-hydroxybutyric acid.

4.3 Amino acids

Some amino acid profiles include GABA in the panel and can be useful for diagnosis of SSADH.

4.4 Example chromatograph

Fig. 11.2.

4.5 Other important diagnostic/monitoring compounds

None applicable.

5 Newborn screening

Succinic semialdehyde dehydrogenase deficiency is not included in the newborn screening programs due to the absence of diagnostic or suggestive biomarkers that can be measured by assays that lend themselves to population screening.

6 Follow-up/confirmatory testing

Diagnosis of SSADH can be established by organic acid analysis for the elevated excretion of 4-hydroxybutyric acid, sometimes accompanied by 4,5-dihydroxyhexanoic acid. Genetic analysis for mutations in the *ALD-H5A1* gene is also available.

7 Interferences and assay or interpretation quirks

Exogenous 4-hydroxybutyrate (gamma-hydroxybutyrate, GHB) may be administered for purposes of sedation or may be present due to it's illicit use as a drug. Exogenous GHB should clear the system and disappear from the urine within 8 hours of intake. Endogenous GHD will remain present in SSADH individuals.

Further reading

[1] NIH Genetic and Rare Diseases Information Center, Succinic Semialdehyde Dehydrogenase Deficiency. https://rarediseases.info.nih.gov/diseases/7695/succinic-semialdehyde-dehydrogenase-deficiency accessed 11/27/2019.
[2] Genetics Home Reference, NIH US National Library of Medicine. https://ghr.nlm.nih.gov/condition/succinic-semialdehyde-dehydrogenase-deficiency accessed 12/6/2019.
[3] GeneReviews, NIH, Succinic Semialdehyde Dehydrogenase deficiency. https://www.ncbi.nlm.nih.gov/books/NBK1195/accessed 11/27/2019.
[4] Gibson KM, Van Hove JLK, Willemsen MAAP, Hoffman GF. Neurotransmitter disorders. In: Sarafoglou K, Hoffman GF, Roth KS, editors. Pediatric endocrinology and inborn errors of metabolism. 2nd ed. New York: McGraw-Hill Co; 2017. p. 1057–91.

SECTION 3

Urea cycle defects

CHAPTER 12

Disorder: Arginase deficiency

1 Synonyms

Argininemia, hyperargininemia.

2 Brief synopsis

2.1 Incidence

~1:100,000–300,000 newborns.

2.2 Etiology

Arginase deficiency (ARG1 deficiency, OMIM # 207800) is a urea cycle disorder.

Arginase 1 (EC3.5.3.1), also known as arginine amidinase, is an enzyme found throughout the body that catalyzes the conversion of arginine to ornithine and urea in the final step of the urea cycle. When activity is deficient and the conversion cannot proceed, arginine concentrations rise as arginine accumulates. (Fig. 12.1) The gene locus for this enzyme is 6q23 and it is an autosomal recessive disorder.

The *ARG1* gene is approximately 10–15 kb in length and pathogenic variants may be found anywhere in the gene. Affected individuals have pathogenic variants on both alleles. The gene encodes a protein of 322 amino acids, which is located in the cytoplasm. Disease outcome and prognosis are more dependent on early diagnosis and appropriate treatment of hyperammonemic crises, than on specific mutations or degree of deficiency.

3 Clinical presentation

Unlike most urea cycle defects, which present with acute hyperammonemia in infancy, arginase deficiency usually presents outside the neonatal period with increasing stiffness and spasticity, especially in the legs. However, most individuals are identified early through expanded newborn screening programs. Plasma ammonia elevations are intermittent and acute hyperammonemia is less common. Individuals with this disorder generally have a slowing of their growth between one and three years of age. This

A Quick Guide to Metabolic Disease Testing Interpretation
http://dx.doi.org/10.1016/B978-0-12-816926-1.00012-2

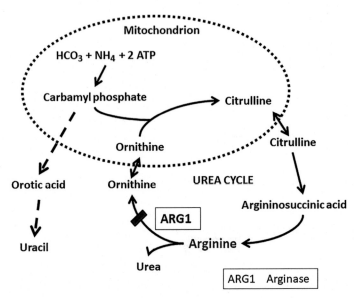

Figure 12.1 *Pathway involving arginase.*

is generally followed by the development of spasticity, and then cognitive development may plateau and be accompanied by a loss of developmental milestones. Seizures, tremor and difficulty with balance and coordination are often seen. Arginase deficiency often progresses to severe spasticity which may result in a loss of ambulation and a loss of bowel and bladder control if not treated. In addition a progressive neurological decline may be seen in untreated or non-compliant cases. In some cases where the enzyme deficiency is less severe, signs and symptoms may appear later in life and also be less severe. Prognosis is mainly dependent on early diagnosis and compliance with treatment regimens. Treatment is to maintain plasma arginine levels as near normal as possible with low protein intake and oral drugs for nitrogen scavenging.

4 Diagnostic compounds

4.1 Urine organic acid profile

Non-diagnostic for arginase deficiency.

4.2 Acylcarnitine profile

Non-diagnostic for arginase deficiency.

4.3 Amino acids

Increased concentration of arginine is the biomarker used to diagnose arginase deficiency. Arginine concentrations are monitored routinely to monitor adherence to treatment in patients diagnosed with arginase deficiency.

4.4 Other important diagnostic/monitoring compounds

Ammonia: Like all urea cycle defects, ammonia concentrations are especially important in monitoring Arginase deficiency. Although hyperammonemia is usually mild in this disdorder, hyperammonemic encephalopathy can occur.

5 Newborn screening

Arginase deficiency is not one of the core disorders in the Newborn Screening Recommended Uniform Screening Panel (RUSP), however it is one of the secondary conditions that are diagnosed in the course of working up the differential on the core conditions. Serum arginine is elevated on amino acid analysis, usually 3–4 times the upper limit of normal.

6 Follow-up/confirmatory testing

Continuous elevation of arginine on serum amino acid analysis or two pathogenic mutations by gene sequencing.

7 Interferences and assay or interpretation quirks

Critically ill patients with hepatic failure can have high plasma arginine concentrations.

Further reading

[1] Vockley JG, Goodman BK, Tabor DE, Kern RM, Jenkinson CP, Grody WW, Cederbaum SD. Loss of function mutations in conserved regions of the human arginase I gene. Biochem Mol Med 1996;59(1):44–51.
[2] Iyer R, Jenkinson CP, Vockley JG, Kern RM, Grody WW, Cederbaum S. The human arginases and arginase deficiency. J Inherit Metab Dis 1998;21(Suppl. 1):86–100.
[3] Genetics Home Reference, NIH US National Library of Medicine. https://ghr.nlm.nih.gov/condition/arginase-deficiency accessed 6/29/2018.
[4] GeneReviews, NIH, Arginase deficiency. https://www.ncbi.nlm.nih.gov/books/NBK1159/accessed 6/29/2018.

[5] Summar ML. Urea cycle defects. In: Sarafoglou K, Hoffman GF, Roth KS, editors. Pediatric endocrinology and inborn errors of metabolism. New York: McGraw-Hill Co; 2009. p. 141–52.
[6] Recommended Uniform Screening Panel. HRSA. https://www.hrsa.gov/advisory-committees/heritable-disorders/rusp/index.html accessed 3/20/2019.
[7] American College of Medical Genetics Newborn Screening Expert G. Newborn screening: toward a uniform screening panel and system–executive summary. Pediatrics 2006;117:S296–307.

CHAPTER 13

Disorder: Argininosuccinic acidemia

1 Synonyms

ASA, argininosuccinic aciduria, ASL deficiency, argininosuccinic acid lyase deficiency, argininosuccinase deficiency

2 Brief synopsis

2.1 Incidence

~1:50,000–70,000 newborns

2.2 Etiology

Argininosuccinate lyase deficiency (ASL deficiency, OMIM # 207900) is a urea cycle disorder. Argininosuccinate lyase (EC4.3.2.1), is the enzyme that performs the fourth step of the urea cycle and converts argininosuccinic acid to arginine and fumarate. Deficiency in this enzyme causes elevated concentrations of argininosuccinic acid and citrulline. Plasma arginine concentrations are often low (Fig. 13.1). The gene locus for this enzyme is 7q11.21 and the disorder is inherited in an autosomal recessive fashion. Pathogenic variants can be found throughout the gene, with sequence analysis detecting pathogenic variants in the coding regions of approximately 90% of individuals who have a biochemical diagnosis of argininosuccinic acidemia (ASL).

3 Clinical presentation

Most urea cycle disorders have a similar severe neonatal onset characterized by acute hyperammonemia in the first few postnatal days. Symptoms include vomiting, lethargy, hypothermia, refusal to feed and respiratory issues, followed by seizures, coma and death if unrecognized and untreated. In addition, in ASL hepatomegaly and coarse, friable hair are often present. In the less severe, late onset form of ASL, hyperammonemia may be episodic and induced by infection, stress or medication non-compliance. ASL has also been associated with an increased incidence of neurocognitive defects

A Quick Guide to Metabolic Disease Testing Interpretation
http://dx.doi.org/10.1016/B978-0-12-816926-1.00013-4

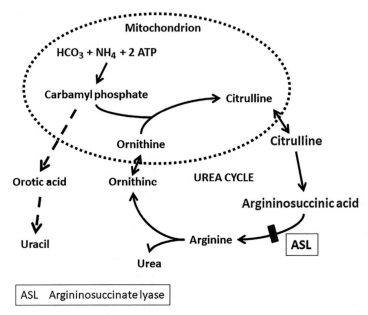

Figure 13.1 *Pathway involving argininosuccinate lyase.*

in comparison to other urea cycle defects. Unlike other urea cycle defects, systemic hypertension may also be seen in ASL.

Like other urea cycle defects, prognosis is dependent on early diagnosis and preventing hyperammonemic episodes. Treatment is to maintain low plasma ammonia concentrations with low protein intake, oral drugs for nitrogen scavenging and arginine supplementation to maintain normal arginine levels, which is essential for the function of the urea cycle.

4 Diagnostic compounds

4.1 Urine organic acid profile

Non-diagnostic for argininosuccinate lyase deficiency although orotic acid may be slightly elevated.

4.2 Acylcarnitine profile

Non-diagnostic for argininosuccinate lyase deficiency

4.3 Amino acids

Plasma: Increased concentration of argininosuccinic acid is the biomarker that distinguishes argininosucciniate lyse deficiency from other urea cycle disorders. Citrulline concentrations are also essentially always elevated in

ASL deficiency. Treatment is monitored with these amino acids as well as monitoring arginine to ensure arginine concentrations do not drop too low. Glutamine concentrations may rise up to 48 hours before ammonia levels go up in ASL making glutamine a surrogate marker for hyperammonemia.

4.4 Other important diagnostic/monitoring compounds

Ammonia: Like all urea cycle defects, ammonia concentrations are especially important in monitoring ASL deficiency. See Chapter 16. (HHH) for a diagnostic scheme for hyperammonemia in urea cycle defects.

5 Newborn screening

ASL is including in the newborn screening programs of all 50 States. Elevated citrulline is used to detect ASA, however since elevated citrulline is also seen with citrullinemia, citrin deficiency and pyruvate carboxylase deficiency, ASL deficiency must be confirmed by plasma amino acid analysis with a finding of elevated argininosuccinic acid. As with most urea cycle disorders, infants usually present with severe hyperammonemia often prior to when the newborn screening results are available. Prompt recognition of these disorders and treatment is necessary to alter the course of the disease and prevent irreversible neurologic damage.

6 Follow-up/confirmatory testing

Diagnosis can be established with amino acid analysis and the findings of elevated argininosuccinc acid in the plasma or urine, plus elevated citrulline and ammonia in the plasma. Mutational analysis of the *ASL* gene can also confirm the diagnosis.

7 Interferences and assay or interpretation quirks

None applicable.

Further reading

[1] GeneReviews, NIH, Argininosuccinate lyase deficiency. https://www.ncbi.nlm.nih.gov/books/NBK51784/accessed 10/15/2018.
[2] Genetics Home Reference, NIH US National Library of Medicine. https://ghr.nlm.nih.gov/condition/argininosuccinic-aciduria# accessed 10/15/2018.
[3] Yudkoff M, Summar ML, Häberle J. Urea cycle disorders. In: Sarafoglou K, Hoffman GF, Roth KS, editors. Pediatric endocrinology and inborn errors of metabolism. 2nd ed. New York: McGraw-Hill Co; 2017. p. 191–208.

CHAPTER 14

Disorder: Carbamyl phosphate synthetase 1 deficiency and N-acetylglutamate synthase deficiency

1 Synonyms

1) carbamoylphosphate synthetase deficiency; CPS1 deficiency, 2) *N*-acetylglutamate synthetase deficiency (NAGSD), NAGS deficiency

2 Brief synopsis

2.1 Incidence

<1:1,000,000 newborns for CPS1 deficiency

   < 1:2,000,000 newborns for NAGS deficiency

2.2 Etiology

Carbamyl phosphate synthetase deficiency (CPS1 deficiency, OMIM # 237300) and *N*-acetylglutamate synthase deficiency (NAGS deficiency, OMIM # 237310) are proximal urea cycle disorders.

Carbamyl phosphate synthetase 1 (EC6.3.4.16), is the rate–limiting enzyme that controls the first step of the urea cycle by converting the ammonia generated by normal catabolic processes into carbamyl phosphate for entry into the urea cycle. The enzyme is almost exclusively found in the liver, inside the mitochondria. When its activity is deficient, ammonia builds up in the body and the urea cycle cannot function normally (Fig. 14.1).

The *CPS1* gene encodes for the enzyme CPS1 and the gene locus for this enzyme is 2q35. The disorder is inherited in an autosomal recessive fashion.

N-acteylglutamine synthase (EC2.3.1.35), also known as glutamate *N*-acetyltransferase, produces the compound *N*-acetylglutamate. *N*-acetylglutamate is necessary for the activity of CPS1 and without it, CPS1 is inactive (Fig. 14.2). Thus signs and symptoms of NAGS deficiency mimic those of CPS1 deficiency. The *NAGS* gene locus for this enzyme is 17q21.31.

A Quick Guide to Metabolic Disease Testing Interpretation
http://dx.doi.org/10.1016/B978-0-12-816926-1.00014-6

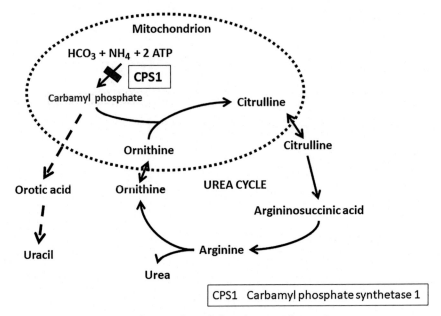

Figure 14.1 *Pathway involving carbamyl phosphate synthetase 1.*

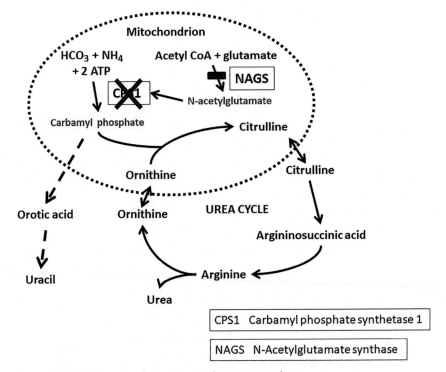

Figure 14.2 *Pathway involving N-acetyglutamate synthase.*

Approximately 12 different mutations have been identified in individuals with NAGS deficiency, and the disorder is inherited in an autosomal recessive manner.

3 Clinical presentation

In general CPS1 and NAGS deficiencies present within the first few days of life with extreme hyperammonemia. Symptoms include sleepiness, respiratory rates and body temperatures that are poorly regulated, unwillingness to feed and vomiting after feeding, lethargy, seizures and coma. These symptoms recur whenever ammonia builds up in individuals who survive the neonatal period, so strict control of diet and strict monitoring of ammonia concentrations are necessary. Individuals with these disorders may have intellectual disabilities and delayed development, and neonatal onset individuals have a poor prognosis. Mild forms of the deficiencies, with partial enzyme activities may present later in life with less severe symptoms. Prognosis is dependent on early diagnosis, prompt treatment of high ammonia concentrations and compliance with treatment regimens. Treatment involves maintaining a low protein diet and oral drugs for nitrogen scavenging. Liver transplant is often the treatment of choice for neonatal onset cases.

4 Diagnostic compounds

4.1 Urine organic acid profile

Non-diagnostic for CPS1 and NAGS deficiencies.

4.2 Acylcarnitine profile

Non-diagnostic for CPS1 and NAGS deficiencies.

4.3 Amino acids

Citrulline and arginine concentrations will be low. Glutamine is often very high, reflecting elevated ammonia concentrations.

4.4 Other important diagnostic/monitoring compounds

Ammonia: Hyperammonemia is the most important laboratory finding, both to direct the differential diagnosis of CPSI deficiency, and for monitoring treatment. There is not a definitive pattern of lab abnormalities for these disorders, and they may be misdiagnosed as neonatal sepsis if ammonia is not ordered.

Urine orotic acid measurement can help differentiate between CPS/ NAGS (not present) and ornithine transcarbamylase deficiency (elevated).

5 Newborn screening

CPS1 and NAGS deficiencies are not among the core disorders in the New-born Screening Recommended Uniform Screening Panel (RUSP), nor are they among the conditions considered to be secondary targets. In general infants with these deficiencies will present before the results of newborn screening are available with episodes of severe hyperammonemia. Measuring ammonia and following an algorithm for the differential diagnosis of hyperammonemic states is usually required. (See Chapter 16 (HHH))

6 Follow-up/confirmatory testing

These conditions are confirmed by mutational analysis of the *CPS1* and/ or *NAGS* genes.

7 Interferences and assay or interpretation quirks

None applicable.

Further reading

[1] Genetics Home Reference, NIH US National Library of Medicine. https://ghr.nlm.nih. gov/condition/carbamoyl-phosphate-synthetase-i-deficiency accessed 6/12/2019.
[2] GeneReviews, NIH, Urea cycle disorders overview. https://www.ncbi.nlm.nih.gov/ books/NBK1217/accessed 6/12/2019.
[3] Yudkoff M, Summar ML, Häberle J. Urea cycle defects. In: Sarafoglou K, Hoffman GF, Roth KS, editors. Pediatric endocrinology and inborn errors of metabolism. 2nd ed. New York: McGraw-Hill Co; 2017. p. 191–208.

CHAPTER 15

Disorder: Citrullinemia and citrin deficiency

1 Synonyms

Citrullinemia type I (CTLNI), argininosuccinate synthase (ASS) deficiency.

Citrullinemia type II (CTLNII), adult-onset citrullinemia, syndromes may include: neonatal intrahepatic cholestasis caused by citrin deficiency (NICCD) and/or failure to thrive and dyslipidemia caused by citrin deficiency (FTTDCD).

2 Brief synopsis

2.1 Incidence

~1:57,000 newborns for Type I,
1:230,000 newborns in Asians for Type II, 1:2,000,0000 outside Asia

2.2 Etiology

Both argininosuccinate synthase deficiency (ASS deficiency or CTLNI, OMIM # 215700) and citrin deficiency (CTLNII, OMIM # 603471, 605814) are urea cycle disorders. Argininosuccinate synthase (EC6.3.4.5) is the enzyme that ligates aspartate and citrulline to form argininosuccinic acid in the urea cycle. Deficiency in this enzyme causes massively elevated concentrations of citrulline and decreased arginine concentrations (Fig. 15.1).

Citrin is a membrane bound aspartate and glutamate carrier protein that transports aspartate out of the mitochondria, making aspartate available for ligation to citrulline in the next step of the urea cycle. Deficiency or dysfunction of this transporter results in decreased aspartate in the cytoplasm, which limits ASS enzyme activity. This results in increased plasma citrulline concentrations (Fig. 15.2).

The gene that codes for argininosuccinate synthase is the *ASS* gene and its location is 9q34.1. Citrin transporter is coded for by the *SLC25A13* gene and its locus is 7q21.3. Both disorders are inherited in an autosomal recessive fashion. There are over 120 known mutations in the *ASS* gene that

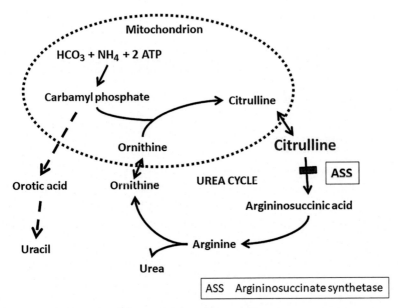

Figure 15.1 *Pathway involving argininosuccinate synthase.*

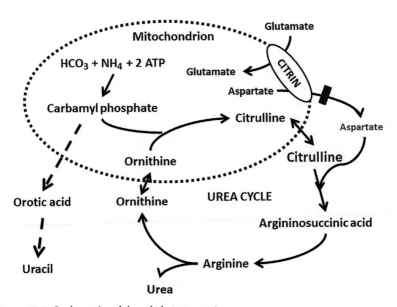

Figure 15.2 *Pathway involving citrin transporter.*

result in citrullinemia type I and at least 20 mutations in the *SLC25A13* gene that result in citrullinemia type II.

3 Clinical presentation

CTLN1 generally presents like most urea cycle defects, in the first few post-natal days of life with acute hyperammonemia. Symptoms are the same as those disorders with vomiting, lethargy, hypothermia, refusal to feed and respiratory issues. Seizures, coma, and death can occur if the hyperammonemia is unrecognized and untreated. CTLN1 can also present in late childhood or adulthood and is associated with a milder phenotype. Symptoms may include intense headaches, blind spots, balance problems and ataxia and lethargy.

Like other urea cycle defects, prognosis is dependent on early diagnosis and preventing episodes of hyperammonemia. The goal of treatment is to maintain low plasma ammonia with low protein intake, nitrogen scavenging agents, and arginine supplementation.

CTLN2 deficiency also has both a neonatal and an adult form. The neonatal form, also called neonatal intrahepatic cholestasis caused by citrin deficiency (NICCD), presents with intrahepatic cholestasis and disturbances in multiple metabolic pathways, including gluconeogenesis, aerobic glycolysis, fatty acid synthesis and catabolism and the urea cycle. Fatty liver and liver fibrosis are common early on, but the signs and symptoms usually go away by a year of age. Affected individuals may develop the early childhood form of CTLN2, also called failure to thrive and dyslipidemia caused by citrin deficiency (FTTDCD). The adult form of CTLN2 is more common that the neonatal form and affects the nervous system, presenting with insidious neurological symptoms such as confusion, irritability, restlessness, abnormal behaviors and memory loss. Patients have brain edema and mortality is high if not treated.

Treatment of CTLN2 is different from other urea cycles disorders, with avoidance of carbohydrate rather than protein. In infants, treatment with formula containing high levels of medium-chain triglycerides has been beneficial. While, the prognosis for the neonatal form of CTLN2 is good, the prognosis for the adult onset form is less clear.

4 Diagnostic compounds

4.1 Urine organic acid profile

Orotic acid can be elevated in ASS deficiency, as in other urea cycle defects (OTC, ASL, ARG1) (Fig. 15.3).

4.2 Acylcarnitine profile

Does not aid in the diagnosis of urea cycle disorders.

4.3 Amino acids

Increased concentration of citrulline is the biomarker used to diagnose citrullinemia. While plasma citrulline concentrations are very high in citrullinemia type I, patients with citrullinemia type II may have milder elevations in plasma citrulline. As with other urea cycle disorders, glutamine concentrations may be used as a surrogate marker for ammonia, as glutamine levels tends to increase before ammonia in plasma.

4.4 Example chromtaograph

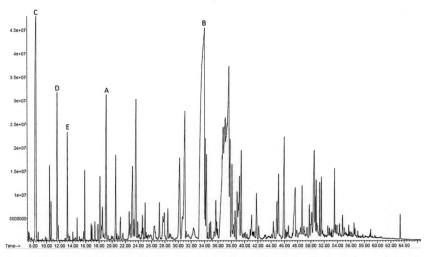

Figure 15.3 *Urine organic acid profile of a patient with CTLN.* (A) Internal standard, (B) orotic acid, (C) lactic acid, and (D–E) ketones (D = 3-hydroxybutyric acid and E = acetoacetic acid).

4.5 Other important diagnostic/monitoring compounds

Ammonia: Severe hyperammonemia is associated with most urea cycle disorders. CTLN II and ARG1 can be associated with milder hyperammoniemia.

5 Newborn screening

CTLN1 is including in the newborn screening programs of all 50 States. Elevated citrulline is used to detect citrullinemia, however since elevated citrulline is also seen with ASL deficiency and pyruvate carboxylase deficiency and CTLN 2, additional testing must be performed to confirm CTLN1, CTLN2 or the other disorders. CTLN2 may not always show an elevated citrulline on newborn screening.

6 Follow-up/confirmatory testing

Diagnosis of CTLN1 can be established with amino acid analysis and the findings of elevated citrulline in the plasma or through mutational analysis of the ASS gene. A diagnosis of citrin deficiency (CTLN2) must be confirmed by assaying for mutations in the *SLC25A13* gene.

7 Interferences and assay or interpretation quirks

None applicable.

Further reading

[1] GeneReviews, NIH, Citrullinemia Type I. https://www.ncbi.nlm.nih.gov/books/NBK1458/ accessed 7/24/2019.
[2] GeneReviews, NIH, Citrin Deficiency. https://www.ncbi.nlm.nih.gov/books/NBK1181/ accessed 7/24/2019.
[3] Genetics Home Reference, NIH US National Library of Medicine. https://ghr.nlm.nih.gov/condition/citrullinemia# accessed 7/24/2019.
[4] Yudkoff M, Summar ML, Häberle J. Urea cycle disorders. In: Sarafoglou K, Hoffman GF, Roth KS, editors. Pediatric endocrinology and inborn errors of metabolism. 2nd ed. New York: McGraw-Hill Co; 2017. p. 191–208.

CHAPTER 16

Disorder: Hyperornithinemia-hyperammonemia-homocitrullinuria syndrome

1 Synonyms

Ornithine translocase deficiency, HHH syndrome, Hyperornithinemia-hyperammonemia-homocitrullinemia syndrome

2 Brief synopsis

2.1 Incidence

Rare; <1:2,000,000
    1:1550 Northern Saskatchewan, CA (founder effect)

2.2 Etiology

Ornithine translocase deficiency, more commonly known as Hyperornithinemia-Hyperammonemia-Homocitrullinuria syndrome, (HHH syndrome, OMIM # 238970) is a urea cycle disorder. Ornithine translocase, also known as mitochondrial ornithine transporter 1 (ORNT1, encoded by *SLC25A15*), is a transporter protein in the mitochondrial membrane that transports ornithine into the mitochondria for use in the urea cycle. Deficiency in this transporter results in insufficient mitochondrial ornithine for ornithine transcarbamylase (OTC) to form citrulline from ornithine and carbamyl phosphate. The accumulated carbamyl phosphate shunts to pyrimidine synthesis resulting in excretion of orotic acid. Additionally, carbamyl phosphate is used in a carbamylation reaction with lysine to form homocitrulline, which is also excreted in the urine. Without a properly functioning urea cycle, hyperammonemia also results (Fig. 16.1).

The gene that codes for ornithine translocase is the *SLC25A15* gene and its location is 13q14.11. There are at least 35 known mutations in the *SLC25A15* gene that result in ornithine translocase deficiency. HHH syndrome is inherited in an autosomal recessive manner, and 30%–50% of affected individuals and most of the French-Canadian population with this disorder carry the Phe188del mutation. The second most common mutation is Arg17Ter, which occurs in 15% of cases.

A Quick Guide to Metabolic Disease Testing Interpretation
http://dx.doi.org/10.1016/B978-0-12-816926-1.00016-X

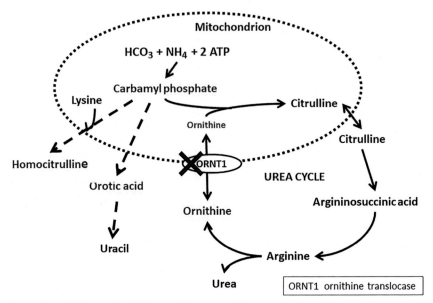

Figure 16.1 *Pathway involving ornithine translocase.*

3 Clinical presentation

HHH may present like other urea cycle defects in the first few post-natal days of life with acute hyperammonemia. Symptoms are the same as those disorders with vomiting, lethargy, hypothermia, refusal to feed and respiratory issues. Seizures, coma, and death can occur if the hyperammonemia is unrecognized and untreated. However, most HHH cases present later and less severely, any time from childhood to adulthood. Symptoms are caused by intermittent hyperammonemia and include lethargy, vomiting, ataxia, spasticity, seizures, vision problems, encephalopathy, developmental delay, and learning disabilities.

Like other urea cycle defects, prognosis is dependent on early diagnosis and preventing hyperammonemia. Therefore, treatment is geared towards restricting protein intake and rigorously treating hyperammonemic episodes. Though this treatment helps to prevent liver dysfunction and hyperammonemic episodes, it does not help with spasticity.

4 Diagnostic compounds

4.1 Urine organic acid profile

Orotic acid excretion is often elevated on urine organic acid analysis. Uracil may also be detected in some urine samples from patients with HHH.

4.2 Acylcarnitine profile

Non-diagnostic for HHH syndrome.

4.3 Amino acids

Urine: The presence of homocitrulline in the urine is used to help diagnose HHH syndrome.

Plasma: Increased concentration of ornithine is helpful although plasma ornithine levels may be normal. Homocitrulline may be found in the plasma as well. Additionally, glutamine will also probably be increased, especially if plasma ammonia concentrations are increased.

4.4 Other important diagnostic/monitoring compounds

Ammonia: Like all urea cycle defects, ammonia concentrations are important in monitoring HHH syndrome. See Fig. 16.2 for a diagnostic scheme for hyperammonemia in urea cycle defects.

Liver dysfunction is often seen with HHH, therefore liver markers may be abnormal, including elevated transaminases and secondary coagulation abnormalities.

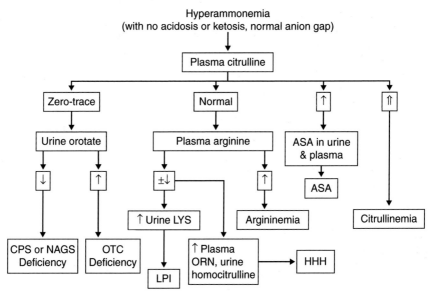

Figure 16.2 *Hyperammonemia diagram for diagnosing urea cycle defects.* *ASA,* ar-gininosuccinic acid; *CPS,* carbamyl phosphate synthetase; *HHH,* hyperonithinemia-hy-perammonemia-homocitrullinuria; *LPI,* lysinuric protein intolerance; *LYS,* lysine; *NAGS, N*-acteylglutamate synthase; *ORN,* ornithine; *OTC,* ornithine transcarbamylase.

5 Newborn screening

HHH syndrome is not included in the newborn screening programs.

6 Follow-up/confirmatory testing

Diagnosis of HHH can be established by amino acid analysis and the findings of persistently elevated ornithine in plasma, homocitrulline in the urine and accompanying episodic or post-prandial hyperammonemia. Genetic testing for the *SLC25A15* gene is available for confirmation.

7 Interferences and assay or interpretation quirks

Homocitrulline may be found in some infant formulas due to carbamylation of lysine during manufacturing. This can cause false positive results. Homocitrulline also co-elutes with methionine in the most commonly performed amino acid analysis (cation exchange chromatography). Thus an elevated urine or plasma methionine may be suggestive of homocitrulline presence when amino acid analysis is performed using this method.

Further reading

[1] GeneReviews, NIH, Hyperonithinemia-Hyperammonemia-Homocitrullinuria Syndrome. https://www.ncbi.nlm.nih.gov/books/NBK97260/ accessed 10/31/2019.
[2] NIH Genetic and Rare Diseases Information Center. Ornithine translocase deficiency. https://rarediseases.info.nih.gov/diseases/2830/hhh-syndrome accessed 10/31/2019.
[3] Genetics Home Reference, NIH US National Library of Medicine. https://ghr.nlm.nih.gov/condition/ornithine-translocase-deficiency# accessed 10/31/2019.
[4] Yudkoff M, Summar ML, Häberle J. Urea cycle disorders. In: Sarafoglou K, Hoffman GF, Roth KS, editors. Pediatric endocrinology and inborn errors of metabolism. 2nd ed. New York: McGraw-Hill Co; 2017. p. 191–208.

CHAPTER 17

Disorder: Ornithine transcarbamylase deficiency

1 Synonyms

OTC

2 Brief synopsis

2.1 Incidence

~1:56,000 newborns

2.2 Etiology

Ornithine transcarbamylase (OTC) deficiency (OMIM #311250) is the most common urea cycle defect. The enzyme catalyzes the condensation of carbamyl phosphate and ornithine to form citrulline. Deficiency of OTC (EC2.1.3.3) results in accumulation of carbamyl phosphate, which enters the cytoplasm and is available for de novo pyrimidine biosynthesis. As a result, orotic acid and uracil will be elevated and excreted into the urine (Fig. 17.1). OTC is an X-linked inborn error of metabolism. The enzyme is a homotrimer encoded by the gene *OTC* that is located on the X chromosome at position Xp11.4. Males tend to have a more severe presentation and worse prognosis. Male infants are at risk of acute metabolic decompensation. Female carriers of the defect may be more or less symptomatic because of skewed X-chromosome inactivation. Any elevated ammonia can result in neurological damage, and even mild deficiencies of OTC can show cumulative damage with repeated episodes of decompensation.

3 Clinical presentation

The primary presenting symptoms of almost all of the urea cycle defects are neurologic, reflecting the neurotoxicity of the hyperammonemia that results from the inability of the liver to form urea from ammonia. OTC often has a severe acute neonatal presentation after the first protein feeding when the ammonia levels increase, especially in male infants who are at risk for subsequent encephalopathy, coma, and death. Symptoms are often difficult to

A Quick Guide to Metabolic Disease Testing Interpretation
http://dx.doi.org/10.1016/B978-0-12-816926-1.00017-1

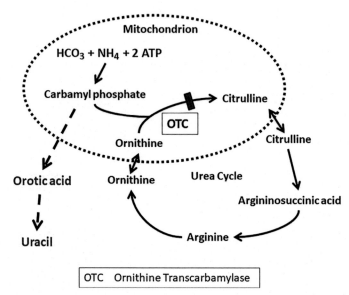

OTC	Ornithine Transcarbamylase

Figure 17.1 *Pathway involving ornithine transcarbamylase.*

differentiate from hypoxia or neonatal sepsis and include increasing lethargy, vomiting, hypothermia, and a respiratory alkalosis. Neurologic posturing may be present if the elevated ammonia has resulted in cerebral edema. Prompt diagnosis may often depend on suspicion of the disorder and measurement of ammonia levels. In later-onset OTC deficiency, the symptoms depend on the extent of the enzyme deficiency and the age of the patient. Generally, acute symptoms are not seen in later onset disorders unless previous episodes have caused brain damage. Symptoms tend to start with vomiting and altered mental status and may include ataxia, confusion, increasing somnolence or irritability, and combativeness. Delayed onset can occur in males and females if the enzyme deficiency is partial or mild. OTC may be difficult to diagnose under these conditions because symptoms may be subtle; for example, episodes of confusion, which can be triggered by such things as a protein challenge or drugs. These patients usually consciously or subconsciously avoid protein in their diet. Delay in making a diagnosis in late-onset OTC deficiency is common.

Like the other urea cycle defects, prognosis is dependent on early diagnosis and keeping ammonia levels down. Treatment is to maintain low plasma ammonia with low protein intake, oral drugs for nitrogen scavenging and citrulline supplementation to help drive the urea cycle. Liver transplant may be the treatment of choice for severe cases.

4 Diagnostic compounds

4.1 Urine organic acid profile

Elevated concentrations of orotic acid and uracil will be present.

4.2 Acylcarnitine profile

Non-diagnostic for OTC deficiency.

4.3 Amino acids

Amino acid analysis is not diagnostic for OTC, however citrulline will be low or absent and arginine may be low as well. Additionally, treatment is monitored by monitoring amino acids as well as ammonia. Glutamine is monitored since concentrations may rise up to 48 hours before ammonia is abnormally elevated, and citrulline may be monitored since it is supplemented in the diet to help drive the urea cycle.

4.4 Example chromatograph

Fig. 17.2, Fig. 17.3 and Fig. 17.4.

4.5 Other important diagnostic/monitoring compounds

Ammonia: Like all urea cycle defects, ammonia concentrations are especially important in monitoring OTC deficiency.

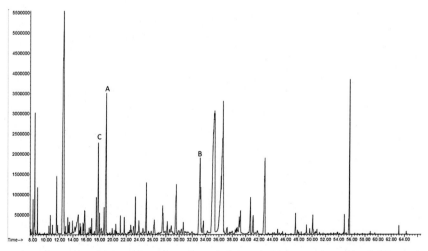

Figure 17.2 *Ornithine transcarbamylase deficiency.* (A) Internal standard, (B) orotic acid, and (C) uracil.

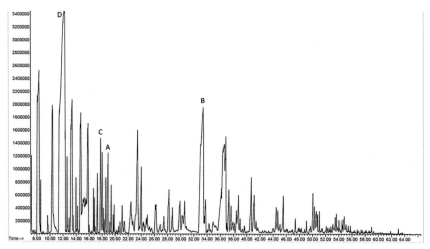

Figure 17.3 *Ornithine transcarbamylase deficiency.* (A) Internal standard, (B) orotic acid, (C) uracil, and (D) large peak of ketones (β-hydroxybutyrate).

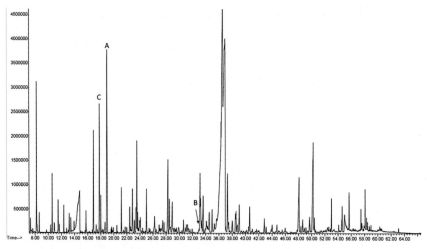

Figure 17.4 *Ornithine transcarbamylase deficiency.* (A) Internal standard, (B) orotic acid (detectable at base of peak), and (C) uracil.

5 Newborn screening

OTC is not including in the newborn screening programs. The two biomarkers used for diagnosis, orotic acid and uracil, are measured via urine organic acid analysis and are not detectable via blood spot analysis by tandem mass spectrometry.

6 Follow-up/confirmatory testing

Diagnosis can be confirmed by molecular testing for pathogenic mutations in the *OTC* gene.

7 Interferences and assay or interpretation quirks

Increased orotic acid and uracil excretion in urine organic acid analysis can be seen in other conditions besides OTC deficiency. Lysinuric protein intolerance, will also show these findings, and orotic acid may be present in urine due to defects in other urea cycle enzymes (citrullinemia type I, arginosuccinate lyase deficiency, and arginase deficiency). A plasma amino acid profile is necessary to differentiate these cases. Additionally, allopurinol treatment can elevate orotic acid excretion in the urine.

Further reading

[1] GeneReviews, NIH, Ornithine transcarbamylase deficiency. https://www.ncbi.nlm.nih. gov/books/NBK154378/ accessed 11/25/2019.

[2] Genetics Home Reference, NIH US National Library of Medicine. https://ghr.nlm.nih. gov/condition/ornithine-transcarbamylase-deficiency# accessed 11/25/2019.

[3] NIH Genetic and Rare Diseases Information Center. Ornithine transcarbamylase deficiency. https://rarediseases.info.nih.gov/diseases/8391/otc-deficiency accessed 11/25/2019.

[4] Yudkoff M, Summar ML, Häberle J. Urea cycle disorders. In: Sarafoglou K, Hoffman GF, Roth KS, editors. Pediatric endocrinology and inborn errors of metabolism. 2nd ed. New York: McGraw-Hill Co; 2017. p. 191–208.

Disorders of amino acid metabolism

CHAPTER 18

Disorder: β-Ketothiolase deficiency

1 Synonyms

BKT, mitochondrial acetoacetyl-CoA thiolase deficiency, MAT, T2 deficiency; mitochondrial 2-methylacetoacetyl-CoA thiolase deficiency, alpha-methylacetoacetic aciduria

2 Brief synopsis
2.1 Incidence

Rare; ~1:1,000,000

2.2 Etiology

β-Ketothiolase deficiency (BKT, OMIM # 203750) is a disorder that involves both the ketone and isoleucine catabolic pathways. β-ketothiolase (EC 2.3.1.16) is also known as mitochondrial acetoacetyl-CoA thiolase (MAT). It both breaks down the ketone acetoacetyl CoA to acetyl-CoA, and also converts 2-methylacetoacetyl-CoA to propionic acid in the isoleucine catabolic pathway. Deficiency in this enzyme impairs the body's ability to process ketones during fatty acid metabolism and as well as impairing the normal breakdown of isoleucine. Excess intermediary metabolites build up, which can be toxic (Fig. 18.1).

The gene that codes for BKT is the *ACAT1* gene and its location is 11q22.3-23.1. More than 70 mutations in the *ACAT1* gene have been identified that result in BKT deficiency, with a few common mutations in specific populations. The disorder is inherited in an autosomal recessive fashion.

3 Clinical presentation

β-ketothiolase deficiency typically presents between 5 months and 4 years of age, often with a metabolic crisis following an intercurrent illness, a period of fasting or increased intake of protein-rich foods. Metabolic acidosis

A Quick Guide to Metabolic Disease Testing Interpretation
http://dx.doi.org/10.1016/B978-0-12-816926-1.00018-3

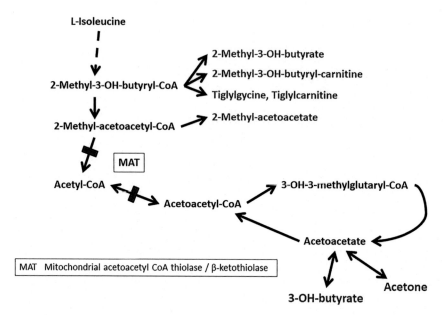

Figure 18.1 *Pathway involving β-ketothiolase (MAT).*

is present, with lethargy and confusion that may proceed to coma. Vomiting, dehydration, and respiratory distress may also be seen. Neurological symptoms have also been reported prior to the first metabolic crisis, including developmental delay. Patients with BKT are generally asymptomatic between metabolic crises.

Treatment of acute episodes involves supplying sufficient glucose to prevent or suppress ketogenesis. Treatment over the long term is generally dietary and includes fasting prevention, avoidance of a fat-rich diet and some protein restriction to keep isoleucine levels down. Prognosis is generally good, with metabolic crises decreasing with age.

4 Diagnostic compounds

4.1 Urine organic acid profile

Urine organic acid analysis will show increased excretion of tiglylglycine, 2-methyl-3-hydroxybutyric acid, and 2-methylacetoacetic acid.

4.2 Acylcarnitine profile

Acylcarnitine profile will show elevated levels of 2-methyl-3-butyrylcarnitine (C5OH-carnitine) and tiglylcarnitine (C5:1-carnitine).

4.3 Amino acids

Non-diagnostic for BKT deficiency.

5 Newborn screening

β-Ketothiolase deficiency is including in the RUSP for newborn screening programs. Elevated concentrations of C5:1- and C5OH-carnitines measured by tandem MS are the biomarkers used for screening for this disorder.

6 Follow-up/confirmatory testing

Because the biomarkers used to screen for and monitor BKT deficiency are seen in other inborn errors of metabolism, confirmation of this disorder should be done with molecular testing.

7 Interferences and assay or interpretation quirks

2-methylacetoacetic acid is unstable and may go undetected in urine organic acid profile. In addition, in mild cases of BKT, the organic acid and acylcarnitine profiles may be normal between metabolic crises, and tiglylglycine may not be detected in urine even during crises.

On acylcarnitine analysis, C5OH-carnitine may also be elevated in 2-methyl-3-hydroxybutyrl-CoA dehydrogenase deficiency, in multiple carboxylase deficiency, in HMG-CoA lyase deficiency and in 3-methylcrotonyl-CoA carboxylase deficiency. Thus elevated C5OH-carnitine is insufficient by itself for diagnosing BKT.

Further reading

[1] NIH Genetic and Rare Diseases Information Center. Ornithine translocase deficiency. https://rarediseases.info.nih.gov/diseases/872/beta-ketothiolase-deficiency   accessed 11/1/2019.
[2] Genetics Home Reference, NIH US National Library of Medicine. https://ghr.nlm.nih.gov/condition/beta-ketothiolase-deficiency# accessed 11/1/2019.
[3] Fukao T, Harding CO. Ketone synthesis and utilization defects. In: Sarafoglou K, Hoffman GF, Roth KS, editors. Pediatric endocrinology and inborn errors of metabolism. 2nd ed. New York: McGraw-Hill Co; 2017. p. 145–59.

CHAPTER 19

Disorder: Lysinuric protein intolerance

1 Synonyms

LPI, congenital lysinuria, dibasic aminoaciduria II, familial protein intolerance

2 Brief synopsis

2.1 Incidence

1:60,000 in Finland, 1:57,000 in Japan, rare elsewhere.

2.2 Etiology

Lysinuric Protein Intolerance (LPI, OMIM # 222700) is caused by a defective transporter protein at the basolateral membrane of kidney and intestinal epithelial cells that does not allow transport of the dibasic/ cationic amino acids, lysine, ornithine and arginine. This transporter is also expressed in hepatic cells, pulmonary epithelial cells and alveolar macrophages. Metabolic abnormalities seen result from deficient absorption, reabsorption and transport of the dibasic amino acids (Fig. 19.1). In LPI, the dibasic amino acids are not absorbed into the blood through the intestinal epithelial cells- and are not reabsorbed from the ultrafiltrate in the renal tubule cells. This results in a shortage of lysine, ornithine and arginine in the body, with diminished concentrations in the blood, and excess amounts of these three amino acids being excreted into the urine.

LPI is caused by mutations in the *SLC7A7* (located on chromosome 14q11.2) gene that encodes for the y + LAT-1 protein which is a subunit of the lysine transporter system that binds the amino acids it is transporting. There are more than 40 known mutations in the *SLC7A7* gene that result in LPI. Common founder variants are known for the Finnish (c.895-2A > T) and Japanese (c.1228C > T) populations. Lysinuric protein intolerance is inherited in an autosomal recessive manner.

A Quick Guide to Metabolic Disease Testing Interpretation
http://dx.doi.org/10.1016/B978-0-12-816926-1.00019-5

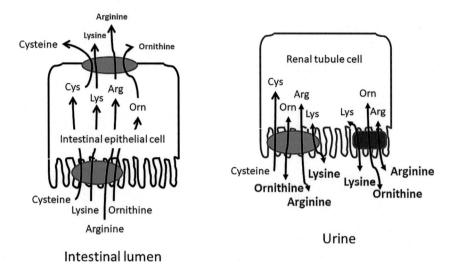

Figure 19.1 *Defective dibasic amino acid transporters in the intestines and kidneys.*

3 Clinical presentation

LPI usually present after weaning since protein concentration in breast milk and formula is low. The initial presentation, with increasing protein in the diet is episodic vomiting and diarrhea, anorexia, decreased growth, hepatosplenomegaly and muscle hypotonia. Infants are at increased risk of acute hyperammonemic coma. This condition originally caused LPI to be thought of as a urea cycle defect. Instead, it appears that low hepatic ornithine and arginine levels impair the urea cycle, resulting in orotic aciduria and increased plasma concentration of glutamine.

LPI is a systemic disorder, with many aspects due to the insufficiency of the essential amino acid lysine. Other systems are increasingly involved as the child gets older. Bone maturation and growth are impaired and pathological fractures before 5 years of age are not uncommon. The hepatomegaly is accompanied by fatty changes of the liver, inflammation and cirrhosis. Pulmonary fibrosis and alveolar proteinosis occur and can result in respiratory insufficiency. Renal glomerular and tubular involvement is common. Anemia, leukopenia and thrombocytopenia are also common, as are high cholesterol, high triglycerides and high ferritin levels.

Treatment involves a protein-restricted diet. Dietary supplementation of the dibasic amino acids is ineffective since they cannot be absorbed. Citrulline supplementation can prevent hyperammonemia and provides arginine and ornithine for the urea cycle. This treatment does not help the pulmonary, renal or skeletal problems.

4 Diagnostic compounds

4.1 Urine organic acid profile

Orotic acid excretion is often elevated on urine organic acid analysis.

4.2 Acylcarnitine profile

Non-diagnostic for LPI.

4.3 Amino acids

Urine: Large amounts of lysine, ornithine and arginine are seen in the urine and are the diagnostic features of LPI.

Plasma: Low concentration of lysine, ornithine and arginine will be present. Glutamine will also be elevated, especially if ammonia levels are elevated. Citrulline is elevated in patients receiving citrulline supplementation.

4.4 Other important diagnostic/monitoring compounds

Ammonia and plasma amino acid levels should be monitored. Deficiency of essential amino acids, especially lysine, is often observed in patients with LPI.

Creatinine and BUN should be monitored for renal involvement and serum concentrations of LD and ferritin are often monitored as well.

5 Newborn screening

LPI is not included in the newborn screening programs.

6 Follow-up/confirmatory testing

LPI should be included in the differential whenever a urea cycle defect is suspected. Diagnosis of LPI can be established by serum amino acid analysis showing low concentrations of the dibasic amino acids, accompanied by urine amino acid showing elevated concentrations of the same, and orotic acid in the urine on organic acid analysis. Confirmation can be done by genetic testing.

7 Interferences and assay or interpretation quirks

At times the lysine, ornithine and arginine seen in the plasma may be low normal rather than outside the reference intervals, and the urine concentrations of these three amino acids may fall within the reference intervals for urine amino acids. It is often useful to calculate the fractional excretion (see calculation below) of these three amino acids. The fractional excretion of most amino acids in the urine is <0.5% because they are predominately reabsorbed by the proximal tubules. In LPI, the fractional excretion of the dibasic amino acids can be between 3% and 75%.

$$\text{Fractional excretion} = [(\text{urine amino acid}/\text{serum amino acid})/$$
$$(\text{urine creatinine}/\text{serum creatinine})] \times 100.$$

Further reading

[1] GeneReviews, NIH, Lysinuric Protein Intolerance. https://www.ncbi.nlm.nih.gov/books/NBK1361/ accessed 11/5/2019.
[2] NIH Genetic and Rare Diseases Information Center. Lysinuric Protein Intolerance. https://rarediseases.info.nih.gov/diseases/3335/lysinuric-protein-intolerance accessed 11/5/2019.
[3] Genetics Home Reference, NIH US National Library of Medicine. https://ghr.nlm.nih.gov/condition/lysinuric-protein-intolerance# accessed 10/31/2019.
[4] Roth KS, Friedman A. Disorders of membrane transport. In: Sarafoglou K, Hoffman GF, Roth KS, editors. Pediatric endocrinology and inborn errors of metabolism. 2nd ed. New York: McGraw-Hill Co; 2017. p. 919–48.

CHAPTER 20

Disorder: Maple syrup urine disease

1 Synonyms

MSUD, branched-chain alpha-ketoacid dehydrogenase deficiency, BCKD deficiency, branched-chain ketoaciduria

2 Brief synopsis

2.1 Incidence

1:185,000; 1:380 in Older Order Mennonite population in Pennsylvania.

2.2 Etiology

Maple Syrup Urine Disease (MSUD, OMIM # 608348, 248611, 248610) is a disorder of branched-chain amino acid (leucine, isoleucine, and valine) metabolism caused by multiple defects in the branched-chain α-ketoacid dehydrogenase multienzyme complex. Branched-chain amino acids (BCAA) comprise approximately 35% of the indispensable amino acids in muscle. They are essential amino acids that are ketogenic and glucogenic, are precursors for fatty acid and cholesterol synthesis, and are substrates for energy production. In the second step in catabolism of the BCAA, branched-chain α-ketoacid dehydrogenase (BCKD; EC1.2.4.4) converts 2-keto-iso-valerate, 2-keto-3-methylvalerate and 2-ketoisocaproate to isobutyrl-CoA, 2-methylbutyryl-CoA and isovaleryl-CoA, respectively (Fig. 20.1). The multienzyme complex of BCKD is comprised of three catalytic enzymes (a decarboxylase, a transacylase, and a dehydrogenase) and two regulatory enzymes (a phosphatase and a kinase) that control the BCKD complex activities. Defects in this complex enzyme system lead to MSUD. Because of this complexity, there is considerable genetic heterogeneity associated with MSUD.

MSUD is inherited in an autosomal recessive manner and is caused by mutations in the *BCKDHA*, *BCKDHB* and *DBT* genes which code for proteins making up the BCKD complex. Mutations in these genes can either decrease or eliminate activity of the complex. *BCKDHA* is located

A Quick Guide to Metabolic Disease Testing Interpretation
http://dx.doi.org/10.1016/B978-0-12-816926-1.00020-1

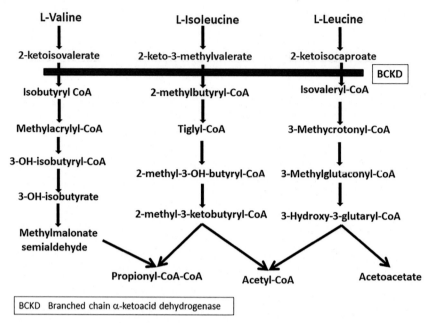

Figure 20.1 *Pathways showing the enzyme defect in MSUD.*

at chromosomal location 19q13.2, with more than 80 mutations identified in individuals with MSUD. *BCKDHB* is located at chromosomal location 6q14.1, with more than 90 mutations identified in MSUD individuals. *DBT* is located at 1p21.2, with more than 70 identified mutations.

3 Clinical presentation

On the basis of clinical and biochemical features, the clinical presentation of MSUD patients is often categorized into four different types. Classic MSUD is the most common and severe phenotype, with BCKD activity less than 2% of normal. It presents with irritability, poor feeding and neonatal-onset progressive neurological deterioration, often without acidosis or other overt signs of metabolic disturbances. Alternating hypertonia and hypotonia is often present. The other forms of MSUD may have up to 30% of normal enzyme activity. Intermediate MSUD presents with slowly progressive failure to thrive, seizures, and developmental delay. Intermittent MSUD presents with ataxia, seizures and coma and often is brought on by an intercurrent illness or a high protein meal. Thiamine-responsive MSUD presents similarly to the intermediate form. The physical signs and symptoms in all cases appear to be related to the accumulation of levels the branched chain amino acids that are toxic to the brain and other organs.

Treatment involves a rapidly decreasing accumulated branched chain amino acids. In addition, formulas free of leucine, isoleucine and valine are often initiated, and life-long avoidance of these amino acids is a staple of treatment. Prognosis is dependent on early diagnosis, rapid treatment, and on good control of BCAA levels.

4 Diagnostic compounds

4.1 Urine organic acid profile

Urine organic acid analysis shows a characteristic pattern of α-keto- and α-hydroxy-derivatives of the branched chain amino acids, including 2-keto and/or 2-hydroxy-isovaleric acid, 2-keto- and/or 2-hydroxy-isocaproic acid, 2-keto- and/or 2-hydroxy-n-caproic acid.

4.2 Acylcarnitine profile

Non-diagnostic for MSUD.

4.3 Amino acids

Plasma: Elevated concentrations of leucine, isoleucine and valine will be present, along with alloisoleucine. Alloisoleucine is pathognomic for MSUD.

4.4 Example chromatograph

Fig. 20.2.

4.5 Other important diagnostic/monitoring compounds

Measurement of BCAA is routinely used to monitor this disorder.

5 Newborn screening

MSUD is included on the RUSP for newborn screening.

Newborn screening for MSUD is based on tandem mass spectrometry (MS/MS) detection of elevated blood levels of leucine, isoleucine, alloiso-leucine, and valine. However, MS/MS as used for NBS without chromato-graphic separation does not distinguish the isobaric amino acids leucine, isoleucine, and alloisoleucine as well as hydroxyproline ("leucines"). The measurement of the ratio of leucines to alanine or the ratio of leucines to phenylalanine increases the sensitivity of the MS/MS screening test.

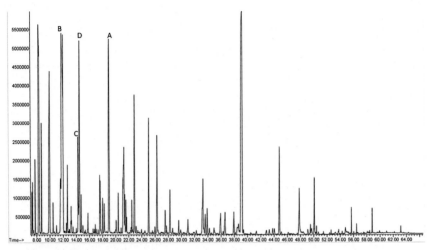

Figure 20.2 *Maple syrup urine disease.* (A) Internal standard, (B) 2-hydroxyisovaleric acid, (C) 2-hydroxyisocaproic acid, and (D) 2-hyroxy-n-caproic acid.

6 Follow-up/confirmatory testing

Confirmatory tests include quantitation of plasma leucine, isoleucine, al-loisoleucine, and valine by amino acid analysis. In particular, alloisoleucine, which is derived from isoleucine, is a sensitive and specific marker of MSUD.

Confirmation can be done by genetic testing if necessary.

7 Interferences and assay or interpretation quirks

With amino acid methods that rely solely on chromatographic separation, plasma alloisoleucine can be falsely elevated due to interference from co-eluting drugs (anti-epileptics). Branched-chain amino acids can also be nonspecifically elevated during starvation or episodes of cyclic vomiting.

2-hydroxyisovaleric, 2-ketoisocaproic, and 2-ketoisovaleric acids can be present in the urine due to extreme ketosis or lactic acidosis; however, these metabolites are not routinely observed in ketotic urine samples.

Further reading

[1] GeneReviews, NIH, Maple Syrup Urine Disease. https://www.ncbi.nlm.nih.gov/books/NBK1319/ accessed 11/8/2019.
[2] NIH Genetic and Rare Diseases Information Center. Maple Syrup Urine Disease. https://rarediseases.info.nih.gov/diseases/3228/maple-syrup-urine-disease   accessed 11/8/2019.

[3] Genetics Home Reference, NIH US National Library of Medicine. https://ghr.nlm.nih. gov/condition/maple-syrup-urine-disease# accessed 11/8/2019.

[4] Hoffmann GF, Burlina A, Barshop BA. Organic acidurias. In: Sarafoglou K, Hoffman GF, Roth KS, editors. Pediatric endocrinology and inborn errors of metabolism. 2nd ed. New York: McGraw-Hill Co; 2017. p. 209–50.

[5] Hong P, Graham KS, Paccou A, Wheat TE, Diehl DM. Compilation of amino acids, drugs, metabolites and other compounds in MassTrak amino acid analysis solution. http://www.waters.com/webassets/cms/library/docs/compilation_of_amino_acids,_ drugs,_metabolites.pdf.

CHAPTER 21

Disorder: Glycine encephalopathy

1 Synonyms

Nonketotic hyperglycinemia, NKH.

2 Brief synopsis

2.1 Incidence

1:60,000

2.2 Etiology

Glycine encephalopathy (OMIM #605899), once called nonketotic hyperglycinemia (NKH), encompasses a number of much more rare disorders. All these disorders have in common the disruption of the glycine cleavage enzyme system (GCS) (Fig. 21.1). The GCS is a multiprotein system found in the mitochondria that is necessary for the cleavage of glycine to donate a methyl group to tetrahydrofolate. Without GCS functioning properly toxic levels of glycine build up in the body and especially in the CNS where glycine acts normally as a neurotransmitter, as well as being pivotal in a number of other reactions in the brain.

The GCS contain 4 protein subunits, P-, T-, H- and *L*-proteins, each encoded for on different genes. The *GLDC* gene encodes for the *P*-protein, is found at location 9p24.1, and mutations in this gene account for roughly 70% of cases of NKH. The *AMT* gene encodes for the T-protein, is found at location 3p21.31, and mutations in *AMT* account for another 25% of cases of NKH. The H-protein is encoded for by the *GCSH* gene, but no patients are known that have a mutation in this gene. The gene for the *L*-protein is currently unknown. NKH is inherited in an autosomal recessive manner.

3 Clinical presentation

NKH is often loosely subdivided on the basis of long-term outcome into severe NKH (85% of cases) and attenuated NKH (15%) of cases. Severe NKH shows essentially no developmental progress and intractable seizures. It presents classically in the first few days of life with lethargy, feeding

A Quick Guide to Metabolic Disease Testing Interpretation
http://dx.doi.org/10.1016/B978-0-12-816926-1.00021-3

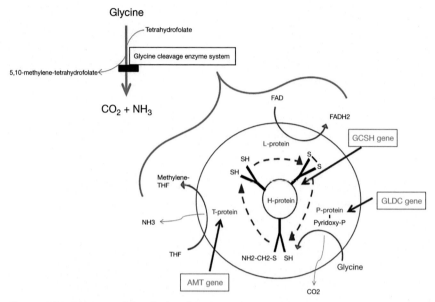

Figure 21.1 *Diagram of the glycine cleavage enzyme system.*

difficulties, marked hypotonia and deep coma, accompanied by neonatal seizure in the form of hiccupping and myoclonic spasms. Mothers of infants with NKH have retrospectively reported feeling the hiccupping in utero. Apnea requiring ventilation support develops usually by the third day of life. Typically, spontaneous breathing returns during the second week of life. Late onset NKH presents later in infancy or childhood and may go on to become classic severe, attenuated or mild. All classic severe cases will lose any developmental progress they may have made, and usually have intractable seizures with no grasping, sitting or contact with their environment. There is no effective treatment, but benzoate to lower glycine levels seems to improve some cases.

Attenuated or mild cases have variable developmental progress and have treatable or even no seizures. Hyperactivity or chorea may be seen in these patients. Treatment involves benzoate to lower glycine levels and using drugs to block the NMDA receptors which are stimulated by glycine. Prognosis in the attenuated cases is highly variable.

4 Diagnostic compounds

4.1 Urine organic acid profile

Non-diagnostic for NKH.

4.2 Acylcarnitine profile

Non-diagnostic for NKH.

4.3 Amino acids

Glycine concentrations are highly elevated in both the plasma and CSF. The ratio of CSF:plasma glycine is over 0.08, where normally it is less than 0.04.

5 Newborn screening

NKH is not included on the RUSP for newborn screening, neither as a core condition nor as a secondary condition.

6 Follow-up/confirmatory testing

Diagnosis of NKH can be made by demonstrating elevated CSF glycine, plasma glycine and an elevated CSF:plasma glycine ratio. Confirmation is usually by molecular testing.

7 Interferences and assay or interpretation quirks

None applicable.

Further reading

[1] NIH Genetic and Rare Diseases Information Center. Glycine Encephalopathy. https://rarediseases.info.nih.gov/diseases/7219/nonketotic-hyperglycinemia accessed 11/22/2019.

[2] Genetics Home Reference, NIH US National Library of Medicine. https://ghr.nlm.nih.gov/condition/glycine-encephalopathy# accessed 11/22/2019.

[3] GeneReviews, NIH, Nonketotic Hyperglycinemia. https://www.ncbi.nlm.nih.gov/books/NBK1357/accessed 11/22/2019.

[4] Gibson KM, Van Hove JLK, Willemsen MAAP, Hoffman GF. Neurotransmitter disorders. In: Sarafoglou K, Hoffman GF, Roth KS, editors. Pediatric endocrinology and inborn errors of metabolism. 2nd ed. New York: McGraw-Hill Co; 2017. p. 1057–91.

CHAPTER 22

Disorder: Phenylketonuria

1 Synonyms

PKU, phenylalanine hydroxylase deficiency, PAH deficiency, hyperphenyl-alaninemia

2 Brief synopsis

2.1 Incidence

~1:10,000–16,500 newborns

2.2 Etiology

Phenylketonuria (PKU, OMIM # 261600) is a disorder of phenylalanine metabolism. Phenylalanine hydroxylase (PAH) (EC1.14.16.1) is the enzyme that converts phenylalanine to tyrosine (Fig. 22.1). Without PAH activity, phenylalanine concentrations build up to toxic levels in the body. The brain is especially sensitive to phenylalanine toxicity and brain damage results from elevated concentrations.

PKU is caused by mutations in the *PAH* gene located at position 12q23.2 online and is inherited in an autosomal recessive manner. Approximately 850 mutations in *PAH* have been identified to date as causative agents of PKU.

PAH activity also requires tetrahydrobiopterin (BH4) as a cofactor for its activity so hyperphenylalaninemia (HPA) is also caused by BH4 deficiency and defects in BH4 metabolism. BH4 metabolic disorders are rare and result from mutations in several genes providing instructions for producing and/or recycling BH4 in the body.

3 Clinical presentation

Individuals with undiagnosed PKU present within the first three months of life with movement disorders, developmental delay, hyper- and hypo-excitability and an abnormal EEG pattern. Often there is an inability to carry out purposeful motions. Retrospectively, the infant is often noted to have been especially irritable with an erratic sleep pattern. Untreated PKU

A Quick Guide to Metabolic Disease Testing Interpretation
http://dx.doi.org/10.1016/B978-0-12-816926-1.00022-5
115

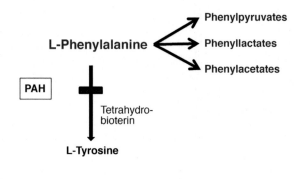

PAH Phenylalanine hydroxylase

Figure 22.1 *Pathway involving phenylalanine hydroxylase.*

individuals often exhibit autistic-like or erratic behavior, aggressiveness, and extreme anxiety or withdrawal. Families are often unable to manage these individuals and they are institutionalized. The disorder progresses to serious to profound mental retardation and abnormal neurological findings.

Treatment of PKU involves keeping phenylalanine concentrations in the body low, with a target range of 120–360 μmol/L. This is especially important during infancy and childhood when elevated phenylalanine concentrations can affect neurological development. Prognosis for a normal life and lifespan is excellent for treated PKU individuals.

4 Diagnostic compounds

4.1 Urine organic acid profile

Phenylpyruvate, phenyllactate, and phenylacetate show increased urine excretion in PKU patients. Mandelate is often also increased. Although PKU is normally diagnosed by measuring plasma phenylalanine concentrations, it can also be detected by the presence of these metabolites in the urine (Fig. 22.2 and Fig. 22.3).

4.2 Acylcarnitine profile

Non-diagnostic for PKU.

4.3 Amino acids

Increased concentration of plasma phenylalanine (PHE) concentrations. In classic PKU, the PHE is usually >1200. PHE concentrations are between

900 and 1200 μmol/L in moderate PKU, ~600–900 μmol/L in mild PKU, and 360–600 μmol/L in hyperphenylalaninemia due to BH4 deficiency or other causes. Tyrosine concentrations may be low.

4.4 Example Chromatographs

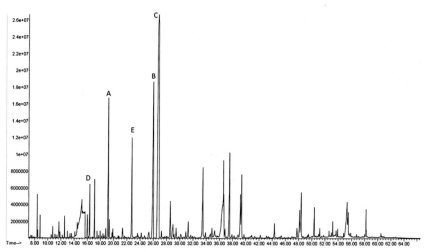

Figure 22.2 *Phenylketonuria.* (A) Internal standard, (B) 2-hydroxyphenylacetic, (C) phenyllactic, (D) phenylacetic, and (E) mandelic acid.

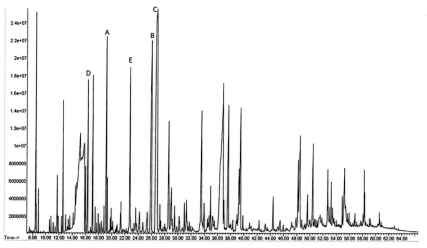

Figure 22.3 *Phenylketonuria.* (A) Internal standard, (B) 2-hydroxyphenylacetic, (C) phenyllactic, (D) phenylacetic, and (E) mandelic acid.

4.5 Other important diagnostic/monitoring compounds

None applicable.

5 Newborn screening

Population testing for PKU was the first newborn screening performed, beginning in Maine in 1960 and using a bacterial inhibition method on dried blood spots for detecting increased phenylalanine concentrations. PKU was the first disorder to be screened for by all 50 states and is included in the RUSP for newborn screening. Elevated phenylalanine concentrations are now detected via LC-MS/MS testing on dried blood spots. Often a ratio of phenylalanine:tyrosine is also examined since phenylalanine concentrations can increase over time with protein load and may still be borderline at the time of newborn screening with milder cases.

6 Follow-up/confirmatory testing

Diagnosis can be established with amino acid analysis and the findings of elevated phenylalanine in the plasma. It is important to also measure urinary pterins (neopterin and biopterin) or enzyme activity of dihydropteridine reductase to separate hyperphenylalaninemia caused by BH4 deficiency from PKU for treatment purposes. Molecular testing for *PAH* mutations can also be done to confirm the diagnosis if necessary.

7 Interferences and assay or interpretation quirks

Excretion of phenylpyruvate, phenyllactate and phenylacetate is not specific for PKU and can be of intestinal bacterial origin (from phenylalanine metabolism). Additionally, phenyllactate and phenylpyruvate can be present secondary to liver diseases or in newborns with hepatic immaturity.

Further reading

[1] GeneReviews, NIH, Phenylalanine Hydroxylase deficiency. https://www.ncbi.nlm.nih. gov/books/NBK1504/accessed 11/26/2019.
[2] Genetics Home Reference, NIH US National Library of Medicine. https://ghr.nlm.nih. gov/condition/phenylketonuria# accessed 11/26/2019.

[3] NIH Genetic and Rare Diseases Information Center. Phenylketonuria. https://raredis-eases.info.nih.gov/diseases/7383/phenylketonuria accessed 11/26/2019.

[4] Burgard P, Luo X, Levy HL, Hoffmann GF. Phenylketonuria. In: Sarafoglou K, Hoffmann GF, Roth KS, editors. Pediatric endocrinology and inborn errors of metabolism. 2nd ed. New York: McGraw-Hill Co; 2017. p. 251–7.

CHAPTER 23

Disorder: Tyrosinemia type 1

1 Synonyms

TYR1, Fumarylacetoacetase deficiency, Fumarylacetoacetate hydrolase (FAH) deficiency, hepatorenal tyrosinosis.

2 Brief synopsis

2.1 Incidence

1:16,000 in Quebec; 1:60,000–74,000 in Sweden; 1:100,000 elsewhere.

2.2 Etiology

Tyrosinemia, type 1 (TYR1) (OMIM #276700) is an autosomal recessive disorder of tyrosine degradation caused by a deficiency in fumarylacetoacetate hydrolase (FAH) (EC3.7.1.2), which results in the accumulation of the enzyme's substrate and precursor, fumarylacetoacetate (FAA) and maleylacetoacetate (MAA), in the liver. In the absence of the enzyme necessary for the metabolism of FAA to fumarate and acetoacetate, these two compounds are converted to succinylacetoacetate (SAA), which is further converted to succinylacetone (SA) (Fig. 23.1). MAA, FAA, SAA, and SA all appear to have biological activity that may be highly toxic and results in progressive tissue damage. MAA and FAA can form glutathione adducts, and these compounds may interfere with glutathione activity. FAA has been shown to have mutagenic activity and may be the causative agent behind the induction of mutations resulting in hepatocellular carcinoma, a known serious complication of TYR1. SA is the metabolite that is detected in the urine and plasma.

TYR1 is caused by mutations in the *FAH* gene located at position 15q23–q25. Approximately 86 mutations in *FAH* have been identified as causative agents of TYR1.

3 Clinical presentation

TYR1 may present at any age, although approximately 90% of patients present in infancy, often within the first 2 months of life. Symptoms are nonspecific and include failure to thrive, irritability, vomiting, and poor

A Quick Guide to Metabolic Disease Testing Interpretation
http://dx.doi.org/10.1016/B978-0-12-816926-1.00023-7

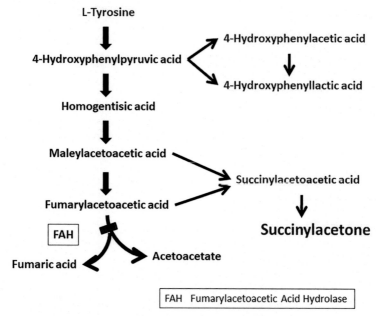

Figure 23.1 *Pathway involving fumarylacetoacetate hydrolase.*

feeding. Hepatic crisis, precipitated by intercurrent infection or other types of metabolic stress, may occasionally be the initial presentation. Some degree of hepatomegaly is usually present, often with splenomegaly, and the abnormal synthetic functions of the liver cause clotting abnormalities. Indications of defects in liver synthetic functions, especially coagulation disturbances, are the most commonly noted liver effects. Jaundice is rare, but it may be present along with ascites and gastrointestinal bleeding. Hepatic crises may resolve or progress to liver failure with hepatic encephalopathy. Along with hepatic disease, a disorder of renal tubular function is usually present with a serious hypophosphatemia. Patients may also have crises resembling porphyria, which is caused by the inhibition of δ-aminolevulinic acid by SA. The most serious complication of Tyr1 is hepatocellular carcinoma with about 18% of patients with TYR1 who live past 2 years old developing hepatocellular carcinoma. The age of onset of symptoms seems to be a good prognostic indicator, with mortality being approximately 60% for infants presenting before 2 months of age and only 4% for individuals presenting after 6 months of age.

Treatment for TYR1 involves the drug nitisonone (2-(2-nitro-4-fluoromethylbenzotyl)-cyclohexane-1,3-dione, also called NTBC) which inhibits 4-hydroxyphenylpyruvate dioxygenase and prevents it from converting 4-hydroxyphenylpyruvate to homogentisic acid. This stops the flux through the tyrosine degradation pathway and toxic metabolites are not formed. Tyrosine concentrations will rise, but succinylacetone will disappear. Symptoms of treated TYR1 individuals will mimic symptoms of Tyrosinemia, type 3, which is a defect in 4-hydroxyphenylpyruvate dioxygenase activity (see Chapter 24, Tyrosinemia 2 and 3). Nitisonone stops acute hepatic and porphyria-like crises, but hepatocellular carcinoma may still develop, especially if nitisonone treatment begins after 2 years of age. Liver transplantation is also a treatment used for TYR1, especially if there is any indication of hepatocellular carcinoma occurring. Routine α-fetoprotein (AFP) measurement is used to monitor TYR1 and progression toward HCC.

4 Diagnostic compounds

4.1 Urine organic acid profile

Succinylacetone in the urine is diagnostic for this disorder. There may also be elevated excretion of other tyrosine metabolites, including 4-hydroxyphenylpyruvic and 4-hydroxyphenyllactic acids.

4.2 Acylcarnitine profile

Does not aid in the diagnosis of TYR1.

4.3 Amino acids

Tyrosine will generally be elevated but not to the extent seen in Tyrosinemias, type 2 and 3, nor transient neonatal hypertyrosinemia. Methionine will usually be elevated as well.

4.4 Example chromatograph

Fig. 23.2.

4.5 Other important diagnostic/monitoring compounds

AFP: AFP is measured routinely to monitor TYR1 and watch for progression to HCC.

Abnormalalities in liver markers are common including elevated transaminases. Additionally, abnormal liver synthesis of clotting factors results in abnormal clotting studies.

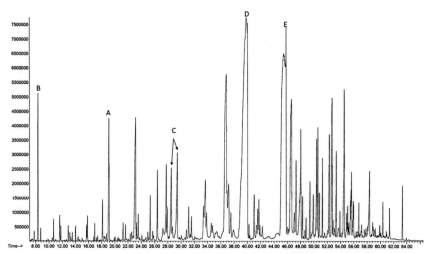

Figure 23.2 *Urine organic acid profile of a patient with TYR1.* (A) Internal standard, (B) lactic acid, (C) two peaks of succinylacetone, (D) 4-hydroxyphenyllactic acid and (E) 4-hydroxyphenylpyruvic.

5 Newborn screening

TYR1 is included in the RUSP for newborn screening. Elevated succinylacetone concentrations are detected in dried blood spots via LCMS/–MS. In TYR1 tyrosine concentrations may be normal on the first newborn screen.

6 Follow-up/confirmatory testing

Diagnosis is usually established with the finding of succcinylacetone in urine organic acid analysis. Amino acid analysis will show increasingly elevated tyrosine concentrations but not to the level of TYR2 and TYR3 until after nitisonone treatment is begun. Molecular testing for *FAH* mutations can also be conducted if necessary.

7 Interferences and assay or interpretation quirks

Transient neonatal hypertyrosinemia can sometimes confound the initial diagnosis of any type of tyrosinemia. TYR1 can fairly rapidly be confirmed due to the presence of SA, which is absent in tyrosinemia types 2 and 3.

Further reading

[1] GeneReviews, NIH, Tyrosinemia, Type 1. https://www.ncbi.nlm.nih.gov/books/NBK1515/accessed 12/6/2019.

[2] Genetics Home Reference, NIH US National Library of Medicine. https://ghr.nlm.nih.gov/condition/tyrosinemia# accessed 12/6/2019.

[3] NIH Genetic and Rare Diseases Information Center. Tyrosinemia, type 1. https://rarediseases.info.nih.gov/diseases/2658/tyrosinemia-type-1 accessed 12/6/2019.

[4] Morris AAM, Chakrapani A. Tyrosinemias and other disorders of tyrosine degradation. In: Sarafoglou K, Hoffmann GF, Roth KS, editors. Pediatric endocrinology and inborn errors of metabolism. 2nd ed. New York: McGraw-Hill Co; 2017. p. 267–86.

CHAPTER 24

Disorder: Tyrosinemia types 2 and 3

1 Synonyms

TYR2, tyrosine aminotransferase deficiency, TAT deficiency, Richner-Hanhart syndrome.

TYR3, 4–hydroxyphenylpyruvate dioxygenase deficiency, HPD deficiency.

2 Brief Synopsis

2.1 Incidence

TYR2: <1:250,000

TYR3: extremely rare, only a few known cases.

2.2 Etiology

Tyrosinemia, type 2 (TYR2) (OMIM #276600) and tyrosinemia, type 3 (TYR3) (OMIM #276710) are disorders of tyrosine degradation caused by deficiencies in tyrosine aminotransferase (TAT) (EC2.6.1.5) and 4–hydroxyphenylpyruvate dioxygenase (HPD) (EC1.13.11.27), respectively. These enzymes catalyze the first two steps in tyrosine degradation, resulting in the accumulation of tyrosine in body tissues (Fig. 24.1).

TYR2 is caused by mutations in the *TAT* gene located at position 16q22.2. There are currently at least 22 known mutations in *TAT* that have been shown to cause TYR2. TYR3 is caused by mutations in the *HPD* gene located at position 12q24.31. Six mutations in *HPD* have been identified as causative agents of TYR3. Both these disorders are inherited in an autosomal recessive manner.

3 Clinical presentation

TYR2 often presents first with ophthalmologic symptoms, including photophobia, pain, redness and lacrimation, frequently within the first year of life. Corneal erosions are usually present. Ophthalmologic

A Quick Guide to Metabolic Disease Testing Interpretation
http://dx.doi.org/10.1016/B978-0-12-816926-1.00024-9

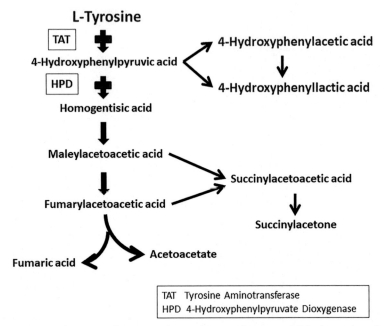

Figure 24.1 *Pathway involving tyrosine aminotransferase and 4-hydroxyphenylpyruvate dioxygenase.*

symptoms occur in roughly 70% of TYR2 patients. Painful hyperkeratotic plaques on the palms of the hands and soles of the feet occur in 80% of TYR2 patients and neurological disabilities occur in approximately 50%. Neurological issues are typically behavioral and cognitive and span a wide range of affects from slight to relatively severe. Many of the symptoms are believed to be caused by tyrosine deposition in tissues due to the very high tyrosine concentrations seen. TYR3 does not have a clearly defined phenotype due to low numbers of diagnosed cases; however, characteristic findings include intellectual disability, developmental delay, seizures and intermitant ataxia. Skin and eye problems are not seen in TYR3.

Treatment for both TYR2 and TYR3 involves diets low in tyrosine and low in phenylalanine, tyrosine's immediate precursor. However, the necessity of a phenylalanine and tyrosine restricted diet in TYR3 patients has not been fully demonstrated. Maintaining lower plasma tyrosine concentrations in TYR2 improves the eye and skin symptoms, however, is not beneficial in improving neurologic and behavioral deficits.

4 Diagnostic compounds

4.1 Urine organic acid profile

TYR2 and TYR3 cannot be differentiated on the basis of urine organic acid analysis. Both disorders will show elevated excretion of tyrosine metabolites, including 4-hydroxyphenylacetic, 4-hydroxyphenylpyruvic and 4-hydroxyphenyllactic acids.

4.2 Acylcarnitine profile

Does not aid in the diagnosis of TYR2 and TYR3.

4.3 Amino acids

Plasma tyrosine concentrations will be markedly increased in Tyrosinemias, type 2 and 3. TYR2 is generally associated with tyrosine concentrations above 1200 μmol/L; however it can be lower in some patients. In TYR3, tyrosine concentrations are usually less than 1200 μmol/L.

4.4 Example chromatograph

Fig. 24.2.

4.5 Other important diagnostic/monitoring compounds

None applicable.

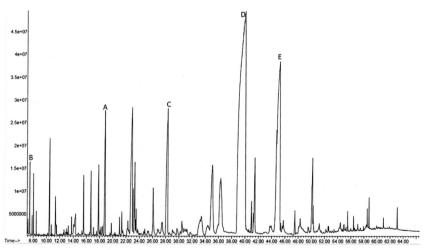

Figure 24.2 *Urine organic acid profile of a patient with TYR2.* (A) Internal standard, (B)-4-hydroxyphenyllactic acid, (C) 4-hydroxyphenylpyruvic and (D) xxxx.

5 Newborn screening

TYR2 and TYR3 are not included in the Core Conditions of the RUSP for newborn screening, however they are considered secondary conditions and will be diagnosed in the differential workup of an elevated tyrosine level on newborn screening, detected by LCMS/–MS amino acid analysis of dried blood spots.

6 Follow-up/confirmatory testing

Diagnosis is suggested by amino acid analysis, which shows continuously elevated tyrosine concentrations accompanied by urine organic acid analysis showing tyrosine metabolites (4–hydroxyphenylacetic, 4–hydroxyphenylpyruvic and 4–hydroxyphenyllactic acids). Diagnostic confirmation is achieved by molecular testing for *TAT* or *HPD* mutations.

7 Interferences and assay or interpretation quirks

Transient neonatal hypertyrosinemia can sometimes confound the initial diagnosis of any type of tyrosinemia. Tyrosinemia types 2 and 3 may require repeated amino acid testing to show continuously elevated levels. TYR2 and TYR3 can be distinguished through clinical features or molecular analysis.

Further reading

[1] Genetics Home Reference, NIH US National Library of Medicine. https://ghr.nlm.nih.gov/condition/tyrosinemia# accessed 12/6/2019.
[2] NIH Genetic and Rare Diseases Information Center. Tyrosinemia, type 2. https://rarediseases.info.nih.gov/diseases/3105/tyrosinemia-type-2 accessed 12/6/2019.
[3] NIH Genetic and Rare Diseases Information Center. Tyrosinemia, type 3. https://rarediseases.info.nih.gov/diseases/10332/tyrosinemia-type-3 accessed 12/6/2019.
[4] Morris AAM, Chakrapani A. Tyrosinemias and other disorders of tyrosine degradation. In: Sarafoglou K, Hoffmann GF, Roth KS, editors. Pediatric endocrinology and inborn errors of metabolism. 2nd ed. New York: McGraw-Hill Co; 2017. p. 267–86.

Fatty acid oxidation defects

CHAPTER 25

Disorder: Carnitine-acylcarnitine translocase deficiency

1 Synonyms

Carnitine-acylcarnitine carrier deficiency (CACT deficiency).

2 Brief synopsis

2.1 Incidence

Very rare; approximately 30 cases reported.

2.2 Etiology

Carnitine-acylcarnitine translocase deficiency (CACT deficiency, OMIM # 212138) is a disorder of the fatty acid uptake and mitochondrial transport system, also known as the carnitine cycle or the carnitine shuttle, which is described in the chapter on carnitine palmitoyl transferase 2, (Chapter 27, CPT2).

Carnitine:acylcarnitine translocase transports long-chain acylcarnitines across the inner mitochondrial membrane and delivers them to CPT2, the next enzyme involved in their transport. (Fig. 25.1) When this CACT is defective or deficient, long chain acylcarnitines are produced by carnitine palmitoyl transferase 1 (CPT1), but cannot be carried into the interior of the mitochondria for oxidation as a fuel source. Reduced energy production results. Long chain acylcarnitines and long chain fatty acids build up and can be toxic to liver, heart and muscles. Free carnitine levels tend to be very low.

The gene which encodes this translocase is the *SLC25A20* gene and its chromosomal location is 3p21.31. Mutations in this gene are heterogeneous and result in structurally abnormal protein and thus a shortage of functional transporter protein. CACT deficiency is inherited in an autosomal recessive fashion.

3 Clinical presentation

CACT is expressed in all tissues requiring fatty acid oxidation (FAO), especially heart, skeletal muscle and liver. CACT deficiency is one of the most severe disorders of the carnitine transport system and mitochondrial

A Quick Guide to Metabolic Disease Testing Interpretation
http://dx.doi.org/10.1016/B978-0-12-816926-1.00025-0

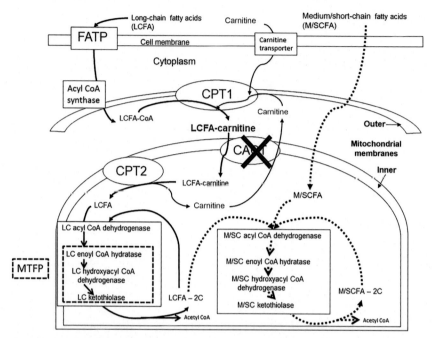

Figure 25.1 *Carnitine cycle pathway involving carnitine:acylcarnitine translocase.* CACT, carnitine:acylcarnitine translocase; CPT1, carnitine palmitoyltransferase 1; CPT2, carnitine palmitoyltransferase 2; FATP, fatty acid transport proteins; MTFP, mitochondrial trifunctional protein.

FAO. Initially presenting as a case of chronic progressive liver failure with hyperammonemia, severe classic presentation is now known to occur at birth with critical illness, including cardiomyopathy with arrhythmias, refractory hyperammonemia, elevated transaminases, severe hypoketotic hypoglycemia, coagulopathies, and elevated creatine kinase. Prognosis is very poor for individuals with the classic presentation. Milder cases with more residual transporter protein available present with moderate myopathy and hepatomegaly. Treatment involves prevention of fasting, low fat diet with fats provided in medium chain lengths, and usually carnitine supplementation.

4 Diagnostic compounds

4.1 Urine organic acid profile

May show a pattern of dicarboxylic aciduria with low or no ketones present. UOA will often be normal when the patient is metabolically stable.

4.2 Acylcarnitine profile

C16- and C18-carnitines will usually be elevated, with low C2-carnitine. Free carnitine is also low, and the acyl fraction is usually elevated.

4.3 Amino acids

Non-diagnostic for CACT deficiency although glutamine may be very high, especially when hyperammonemia is present.

4.4 Other important diagnostic/monitoring compounds

Ammonia: Hyperammonemia is important to monitor as it tends to be refractory to treatment and must be kept low.

Creatine Kinase: CK is usually elevated in CACT deficiency. Transaminases are also elevated and glucose is often low.

5 Newborn screening

CACT deficiency is not one of the core disorders in the Newborn Screening Recommended Uniform Screening Panel (RUSP), however it is a secondary target that should be detected during the differential diagnosis of the core conditions. The biomarkers that indicate the possibility of CACT deficiency in newborn screening are elevated C16-, C18-, C18:1- and C18:2-carnitines. Other biomarkers that may also be added include C14- and C16:1-carnitines and a low ratio of free carnitine to combined C16 and C18 (C0/[C16 + C18]).

6 Follow-up/confirmatory testing

CACT deficiency generally requires full gene sequencing of the *SLC25A20* gene, followed by analysis of the deletions and duplications for confirmation of diagnosis.

7 Interferences and assay or interpretation quirks

All mitochondrial FAO and carnitine transport system deficiencies which involve the long-chain fatty acid species have acylcarnitines profiles that can look alike, with elevations ranging from C14- through C18-species. Looking for the predominant elevations can help interpret to profile, as can looking at the clinical picture and other lab tests results.

Further reading

[1] Genetics Home Reference, NIH US National Library of Medicine. https://ghr.nlm.nih.gov/condition/carnitine-acylcarnitine-translocase-deficiency#genes   accessed 6/14/2019.
[2] Vockley J, Longo N, Andresen BS, Bennett MJ. Mitochondrial fatty acid oxidation defects. In: Sarafoglou K, Hoffman GF, Roth KS, editors. Pediatric endocrinology and inborn errors of metabolism. 2nd ed. New York: McGraw-Hill Co; 2017. p. 125–44.
[3] Recommended Uniform Screening Panel. HRSA. https://www.hrsa.gov/advisory-committees/heritable-disorders/rusp/index.html accessed 3/20/2019.
[4] American College of Medical Genetics Newborn Screening Expert G. Newborn screening: toward a uniform screening panel and system--executive summary. Pediatrics 2006;117:S296-307.

CHAPTER 26

Disorder: Carnitine palmitoyltransferase 1 deficiency

1 Synonyms

Carnitine palmitoyltransferase I deficiency, carnitine palmitoyltransferase 1A deficiency.

2 Brief synopsis

2.1 Incidence

Very rare; <50 cases reported.

2.2 Etiology

Carnitine palmitoyltransferase 1A deficiency (CPT1A deficiency, OMIM # 255120) is a disorder of the fatty acid uptake and mitochondrial transport system, also known as the carnitine cycle or the carnitine shuttle, which is described in the chapter on carnitine palmitoyl transferase 2 (Chapter 27, CPT2).

Carnitine palmitoyltransferase 1A (EC 2.3.1.21) is embedded in the outer mitochondrial membrane and is the rate-limiting enzyme for transporting long chain fatty acids into the mitochondria. CPT1A generates long-chain acylcarnitines from long chain acyl-CoAs and free carnitine, in preparation for their transport across the inner mitochondrial membrane by CACT. (Fig. 26.1) CPT1A is expressed in the liver, and its deficiency leads to the inability of long–chain fatty acids to enter to mitochondria to be used as a source of energy and ketones. Long chain fatty acids build up, especially in the liver. Free carnitine levels are generally high. CPT1B is expressed in skeletal and cardiac muscle and CPT1C is expressed in brain tissue. Disorders in CPT1B and CPT1C have not been described.

The gene which encodes CPT1A is the *CPT1A* gene and its chromosomal location is 11q13.1. More than 20 mutations in the *CPT1A* gene have been shown to severely reduce or eliminate CPT1A activity and cause this disorder. CPT1A deficiency is inherited in an autosomal recessive fashion.

A Quick Guide to Metabolic Disease Testing Interpretation
http://dx.doi.org/10.1016/B978-0-12-816926-1.00026-2
137

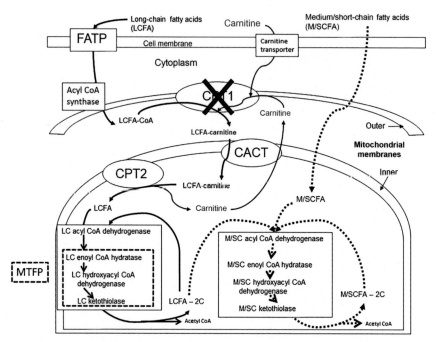

Figure 26.1 *Carnitine cycle pathway involving carnitine palmitoyltransferase 1A.* CACT, carnitine:acylcarnitine translocase; *CPT1*, carnitine palmitoyltransferase 1; *CPT2*, carnitine palmitoyltransferase 2; *FATP*, fatty acid transport proteins; *MTFP*, mitochondrial trifunctional protein.

3 Clinical presentation

CPT1A is expressed predominately in liver tissue, so clinical presentation and symptoms are related to liver involvement. The disorder usually presents within the first year of life, with the presentation often induced by fasting. The patients present with hepatic encephalopathy, hyperammonemia, hypoglycemia, elevated transaminases, low ketones and coagulopathies. Lethargy is usually present and will progress to coma without appropriate treatment. CPT1A is not expressed in cardiac or skeletal muscle, thus unlike defects in other carnitine shuttle and FAO enzymes, there is generally no muscle involvement.

Treatment of CPT1A deficiency involves prevention of fasting, a low fat diet with provision of fats for energy as medium–chain fatty acids in the form of medium-chain triglycerides. Medium chain length fatty acids do not require the carnitine shuttle to be able to enter the mitochondria. Prognosis seems to be dependent upon the efficacy of prevention of catabolic states during infancy.

4 Diagnostic compounds

4.1 Urine organic acid profile

May show a pattern of medium chain dicarboxylic aciduria. This is a non-specific finding. UOA will often be normal when the patient is metabolically stable.

4.2 Acylcarnitine profile

Total carnitine is often elevated, with the majority or all of the carnitine being in the free form. Long-chain acylcarnitines (C16- and C18-) will usually be very low.

4.3 Amino acids

Non-diagnostic for CPT1A deficiency although glutamine may be very high, especially when hyperammonemia is present.

4.4 Other important diagnostic/monitoring compounds

Ammonia: Hyperammonemia is important to monitor in all FAO disorders.

5 Newborn screening

CPT1A deficiency is not one of the core disorders in the Newborn Screening Recommended Uniform Screening Panel (RUSP), however it is a secondary target that should be detected during the differential diagnosis of the core conditions. The biomarkers used to indicate CPT1A deficiency in newborn screening include very high free carnitine concentrations with low C16- and C18-carnitines. The ratio of free carnitine to combined C16 and C18 (C0/[C16 + C18]) is also often used and will be very high.

6 Follow-up/confirmatory testing

CPT1A deficiency generally requires full gene sequencing of the *CPT1A* gene for confirmatory diagnosis.

7 Interferences and assay or interpretation quirks

All mitochondrial FAO and carnitine transport system deficiencies which involve the long-chain fatty acid species have acylcarnitines profiles that can look alike, with elevations ranging from C14- through C18-species.

Looking for the predominant elevations can help interpret to profile, as can looking at the clinical picture and other lab tests results. In CPT1A deficiency, very elevated concentrations of free carnitine can be a valuable clue.

Further reading

[1] Genetics Home Reference, NIH US National Library of Medicine. https://ghr.nlm. nih.gov/condition/carnitine-palmitoyltransferase-i-deficiency#inheritance   accessed 6/18/2019.

[2] Vockley J, Longo N, Andresen BS, Bennett MJ. Mitochondrial fatty acid oxidation defects. In: Sarafoglou K, Hoffman GF, Roth KS, editors. Pediatric endocrinology and inborn errors of metabolism. 2nd ed. New York: McGraw-Hill Co; 2017. p. 125–44.

[3] Recommended Uniform Screening Panel. HRSA. https://www.hrsa.gov/advisory-committees/heritable-disorders/rusp/index.html accessed 3/20/2019.

[4] American College of Medical Genetics Newborn Screening Expert G. Newborn screening: toward a uniform screening panel and system--executive summary. Pediatrics 2006;117:S296-307.

CHAPTER 27

Disorder: Carnitine palmitoyltransferase 2 deficiency

1 Synonyms

Carnitine palmitoyltransferase II deficiency, (CPT2 deficiency).

2 Brief synopsis

2.1 Incidence

Rare; reported cases include >300 cases of the myopathic form, <20 lethal neonatal form, ~30 severe infantile hepatocardiomuscular form.

2.2 Etiology

Carnitine palmitoyltransferase 2 deficiency (CPT2 deficiency, OMIM # 608836) is a disorder of the fatty acid uptake and mitochondrial transport system, also known as the carnitine cycle or the carnitine shuttle.

The carnitine shuttle involves (1) cellular uptake of the long-chain fatty acids by fatty acid transport proteins and carnitine by a carnitine transporter in the cell membrane; (2) the activation of long-chain fatty acids to acyl-CoA esters by the outer mitochondrial membrane enzyme acyl-CoA synthase; (3) transfer of long-chain fatty acids across the inner mitochondrial membrane by the carnitine shuttle involving formation of acylcarnitines by carnitine palmitoyl transferase 1 (CPT1), transport of acylcarnitines and carnitine across the inner mitochondrial membrane by carnitine/acylcarnitine translocase (CACT), and re-formation of acyl-CoA esters by carnitine palmitoyl transferase 2 (CPT2) embedded in the inner mitochondrial membrane. Disorders of the other constituents of the carnitine shuttle are discussed in their own chapters. Carnitine palmitoyltransferase 2 (CPT2) converts the transported long-chain ac-ylcarnitines back to long-chain acyl-CoAs for entry into the fatty acid oxidation pathway. CPT2 also releases free carnitine for re-use in the carnitine shuttle (Fig. 27.1). Like CACT, when CPT2 is deficient or absent long-chain acylcarnitines build up and cannot be utilized by the mitochondrial fatty acid oxidation pathway for energy, which is espe-

A Quick Guide to Metabolic Disease Testing Interpretation
http://dx.doi.org/10.1016/B978-0-12-816926-1.00027-4

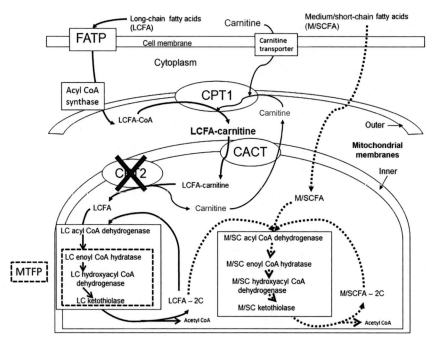

Figure 27.1 *Carnitine cycle pathway involving carnitine palmitoyltransferase 2. CACT,* carnitine:acylcarnitine translocase; *CPT1,* carnitine palmitoyltransferase 1; *CPT2,* carnitine palmitoyltransferase 2; *FATP,* fatty acid transport proteins; *MTFP,* mitochondrial trifunctional protein.

cially critical during fasting or times of heavy energy expenditure. Free carnitine levels tend to be low. CPT2 is expressed in the liver, skeletal and cardiac muscle, and the disorder presents as three main types of CPT2 deficiency discussed below.

The *CPT2* gene encodes the protein carnitine palmitoyltransferase 2 and mutations in this gene give rise to CPT2 deficiency. Its chromosomal location is 1p32.3. More than 70 mutations in the *CPT2* gene have been described, leading to reduced or absent enzyme activity. There appears to be some genotype:phenotype correlation in CPT2 deficiency. The most common *CPT2* gene mutation substitutes a leucine for a serine at position 113 (Ser113Leu or S113L) of the protein. This missense mutation accounts for approximately 60% of the cases of the myopathic form of CPT2 deficiency. *CPT2* null mutations which truncate the protein or cause mRNA degradation lead to the lethal neonatal form of CPT2 deficiency. CPT2 deficiency is inherited in an autosomal recessive fashion.

3 Clinical presentation

CPT2 deficiency has three distinct clinical presentations. The most severe is the lethal neonatal form which present within days of birth with hypoketotic hypoglycemia and liver failure. Cardiomyopathy is present, with cardiac arrhythmias. Fasting or intercurrent infection often leads rapidly to seizures and coma. Facial or structural abnormalities may be present and cystic dysplasia of the kidneys and brain is common. This form has a very poor prognosis and is often rapidly fatal within days to months.

The second form of CPT2 deficiency, the severe infantile hepatocardiomuscular form, is slightly less severe in that it presents later, although usually still within the first year of life. Symptoms and biochemical findings are similar to the lethal neonatal form, with the addition of peripheral myopathy. Attacks of abdominal pain and headaches are also common.

The myopathic form of CPT2 deficiency may be mild. It is considered the most common disorder of lipid metabolism affecting skeletal muscle. Its onset occurs generally anywhere from 10 to 60 years old, and occurs episodically as recurrent myalgia attacks that may develop into rhabdomyolysis. Myoglobinuria accompanying these attacks can lead to renal failure. These attacks are most commonly precipitated by exercise, followed by infection and then fasting. Exposure to cold, sleep deprivation, stress and general anesthesia can also trigger attacks. Creatine kinase concentrations will be massively elevated during attacks but are usually normal between attacks.

Treatment of CPT2 deficiency involves prevention of fasting and a low fat diet with provision of fats for energy as medium-chain fatty acids in the form of medium chain triglycerides. Medium chain length fatty acids do not require the carnitine shuttle to be able to enter the mitochondria. In the case of the myopathic form, avoidance of triggers is important.

4 Diagnostic compounds

4.1 Urine organic acid profile

May show a pattern of medium chain dicarboxylic aciduria. This is a nonspecific finding. UOA will often be normal when the patient is metabolically stable.

4.2 Acylcarnitine profile

Free carnitine is often low, and total carnitine may be low also. Long-chain acylcarnitines, especially C16- and C18:1- carnitines will be elevated.

4.3 Amino acids

Non–diagnostic for CPT2 deficiency.

4.4 Other important diagnostic/monitoring compounds

Ammonia: Hyperammonemia is important to monitor in all FAO disorders.

Creatine kinase: CK will be elevated during attacks in the myopathic form and may be elevated at times in the severe infantile hepatocardiomuscular form.

5 Newborn screening

CPT2 deficiency is not one of the core disorders in the Newborn Screening Recommended Uniform Screening Panel (RUSP), however it is a secondary target that should be detected during the differential diagnosis of the core conditions. The biomarkers used to indicate CPT2 deficiency in newborn screening include low free carnitine concentrations and elevated long chain acylcarnitines species, especially C16- and C18-carnitines. The pattern is similar to that seen with CACT deficiency, with also a low ratio of free carnitine to combined C16 and C18 (C0/[C16 + C18]). Serum total carnitine is usually normal.

6 Follow-up/confirmatory testing

Because the pattern of biochemical markers seen in CPT2 deficiency is so similar to that of CACT, genetic testing is necessary for confirmation of this disorder.

7 Interferences and assay or interpretation quirks

As previously mentioned, all mitochondrial FAO and carnitine transport system deficiencies, which involve the long-chain fatty acid species have acylcarnitines profiles that can look alike, with elevations ranging from C14- through C18-species. Looking for the predominant elevations can help interpret to profile, as can looking at the clinical picture and other lab tests results.

Further reading

[1] Genetics Home Reference, NIH US National Library of Medicine. https://ghr.nlm.nih.gov/condition/carnitine-palmitoyltransferase-ii-deficiency# accessed 6/25/2019.

[2] Vockley J, Longo N, Andresen BS, Bennett MJ. Mitochondrial fatty acid oxidation defects. In: Sarafoglou K, Hoffman GF, Roth KS, editors. Pediatric endocrinology and inborn errors of metabolism. 2nd ed. New York: McGraw-Hill Co; 2017. p. 125–44.

[3] GeneReviews, NIH, Carnitine Palmitoyltransferase II deficiency. https://www.ncbi.nlm.nih.gov/books/NBK1253/accessed 6/26/2019.

[4] Recommended Uniform Screening Panel. HRSA. https://www.hrsa.gov/advisory-committees/heritable-disorders/rusp/index.html accessed 3/20/2019.

[5] American College of Medical Genetics Newborn Screening Expert G. Newborn screening: toward a uniform screening panel and system--executive summary. Pediatrics z2006;117:S296-307.

CHAPTER 28

Disorder: Carnitine transporter deficiency

1 Synonyms

Carnitine uptake defect (CUD), systemic primary carnitine deficiency (CDSP).

2 Brief synopsis

2.1 Incidence

1:22,000–77,000

2.2 Etiology

Carnitine transporter deficiency (CUD, OMIM # 212140) is a disorder of the fatty acid uptake and mitochondrial transport system, also known as the carnitine cycle or the carnitine shuttle, which is described in the chapter on carnitine palmitoyltransferase 2 (CPT2).

Carnitine in transported into the cells via a high affinity transporter called OCTN2. This is a sodium-dependent organic cation transporter, which is located in the plasma membrane of skeletal and cardiac muscle and the renal tubules. When it is deficient, carnitine is lost in the urine, and muscle cells are unable to take carnitine up and concentrate it within the cells. The lack of carnitine in the cells impairs the carnitine cycle and thus fatty acid oxidation (FAO) (See Fig. 28.1).

The *SLC25A5* gene encodes the OCTN2 carnitine transporter and the gene is located on chromosome 5q31.1. More than 60 mutations in the *SLC25A5* gene are known to result in absent or dysfunctional OCTN2 transporter protein and cause CUD. Carnitine transporter deficiency is inherited in an autosomal recessive fashion.

3 Clinical presentation

CUD has an excessively variable presentation, ranging from sudden death in infancy after metabolic decompensation, to asymptomatic adults. Symptoms

A Quick Guide to Metabolic Disease Testing Interpretation
http://dx.doi.org/10.1016/B978-0-12-816926-1.00028-6

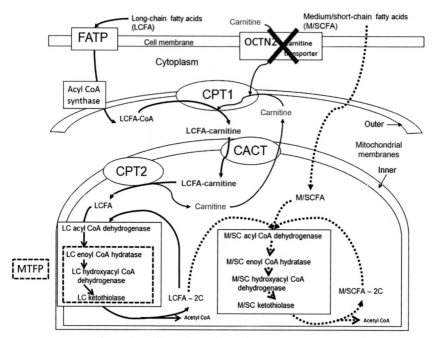

Figure 28.1 *Carnitine cycle pathway involving carnitine transporter OCTN2.* OCTN2, organic cation carnitine transporter; *CACT*, carnitine:acylcarnitine translocase; *CPT1*, carnitine palmitoyltransferase 1; *CPT2*, carnitine palmitoyltransferase 2; *FATP*, fatty acid transport proteins; *MTFP*, mitochondrial trifunctional protein.

appear in infancy after carnitine from the mother has been depleted and they are often precipitated by fasting or fever. Symptoms include encephalopathy, cardiomyopathy, hypoglycemia, vomiting, and muscle weakness.

The liver utilizes a different carnitine transporter than the OCTN2 transporter. The liver transporter functions in CUD however the lack of carnitine in general still results in the inability to get fatty acids into the liver mitochondria for FAO.

Older pediatric patients generally present with a progressive cardiomyopathy, which is fatal if untreated. Adults with CUD may be asymptomatic or may show fatigability.

Treatment for CUD involves high dose carnitine supplementation, which has been shown to reverse the cardiomyopathy caused by the disease. Prognosis of treated individuals is very good, but repeated attacks of hypoglycemia or death from arrhythmias have been reported if carnitine treatment is not continued.

4 Diagnostic compounds

4.1 Urine organic acid profile

Like many defects in FAO or FAO-related pathways, this analysis may show a pattern of medium chain dicarboxylic aciduria. This is a non-specific finding. UOA will often be normal when the patient is metabolically stable.

4.2 Acylcarnitine profile

Free and total carnitine concentrations are low. Acylcarnitine species in general are all low.

4.3 Amino acids

Non-diagnostic for CUD.

4.4 Other important diagnostic/monitoring compounds

Total and free carnitine: These concentrations should be monitored at least twice a year in children, and annually in adults.

Creatine kinase: CK may be elevated during decompensations.

5 Newborn screening

Carnitine transporter deficiency is one of the core disorders in the Newborn Screening Recommended Uniform Screening Panel (RUSP). The biomarkers used to indicate CUD in newborn screening include low free and total carnitines.

6 Follow-up/confirmatory testing

Confirmatory testing for CUD generally requires sequencing of the entire coding regions to look for two pathogenic mutations, as most currently known mutations are private.

7 Interferences and assay or interpretation quirks

Because of overall low concentrations of carnitine, acylcarnitine analysis may not be informative, with no elevated concentrations of any species.

Further reading

[1] Genetics Home Reference, NIH US National Library of Medicine. https://ghr.nlm.nih.gov/condition/primary-carnitine-deficiency# accessed 7/19/2019.

[2] Vockley J, Longo N, Andresen BS, Bennett MJ. Mitochondrial fatty acid oxidation defects. In: Sarafoglou K, Hoffman GF, Roth KS, editors. Pediatric endocrinology and inborn errors of metabolism. 2nd ed. New York: McGraw-Hill Co; 2017. p. 125–44.

[3] GeneReviews, NIH, Systemic Primary Carnitine Deficiency. https://www.ncbi.nlm.nih.gov/books/NBK84551/accessed 7/19/2019.

CHAPTER 29

Disorder: Long chain 3-hydroxyacyl-CoA dehydrogenase deficiency

1 Synonyms

LCHAD deficiency, trifunctional protein deficiency, type 1, TFP deficiency.

2 Brief synopsis

2.1 Incidence

1:250,000–750,000; 1:62,000 in Finnish population.

2.2 Etiology

Long chain 3-hydroxyacyl-CoA dehydrogenase deficiency (LCHAD deficiency, OMIM # 609016) is a disorder of the mitochondrial fatty acid β-oxidation pathway. LCHAD (EC 1.1.1.211) is one of three enzymes which make up the trifunctional protein embedded in the inner mitochondrial membrane. These are three of the four enzymes responsible for cycling long chain fatty acids through β-oxidation and shortening those fatty acids by a 2-carbon acetyl group with each passage through the cycle. The breakdown of long chain fatty acids is a requirement for normal functioning of high-energy requiring tissues like heart and skeletal muscle, but is also necessary for all tissues during fasting. Both trifunctional protein deficiency, which shows a generalized decrease in activity in all three enzymes, and LCHAD deficiency impair the breakdown of long chain fatty acids for energy and result in the abnormal build-up of long chain fatty acid intermediates, which appear to be toxic to liver, heart, muscles and retina (Fig. 29.1).

The *HADHA* gene encodes the LCHAD enzyme part of the trifunctional protein. TFP has 8 subunits. Four alpha subunits are encoded by the *HADHA* gene and include LCHAD and long chain 2-enoyl-CoA hydratase enzyme activities. The four beta subunits are encoded by the *HADHB* gene and contain the third enzyme, long chain ketothiolase. Mutations in the *HADHA* gene give rise to LCHAD deficiency and are inherited in an

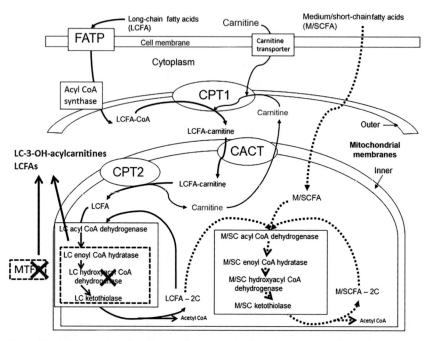

Figure 29.1 *Fatty acid oxidation pathway involving long chain 3-hydroxyacyl-CoA dehydrogenase.* *CACT*, carnitine:acylcarnitine translocase; *CPT1*, carnitine palmitoyltransferase 1; *CPT2*, carnitine palmitoyltransferase 2; *FATP*, fatty acid transport proteins; *MTFP*, mitochondrial trifunctional protein.

autosomal recessive fashion. Its chromosomal location is 2p.23.3. Most cases of LCHAD deficiency are due to a common mutation of c.1528 G > C, p.E474N in the mature protein subunit.

3 Clinical presentation

LCHAD deficiency typically presents in infancy or early childhood with feeding difficulties, lethargy, hypoketotic hypoglycemia and general liver dysfunction. Presentation may be as a Reye-like syndrome. These individuals are also at risk for serious cardiomyopathy, breathing difficulties, coma and sudden death. Uniquely among the fatty acid oxidation disorders, roughly 70% of individuals with LCHAD disorder develop chorioretinal atrophy with loss of visual acuity. Individuals may also develop severe muscle pain and weakness with exercise or exposure to cold and some develop peripheral neuropathies. These muscle symptoms seem to predominate as the LCHAD deficient individual gets older.

Also uniquely, mothers carrying LCHAD deficient fetuses are prone to liver disease, specifically acute fatty liver of pregnancy (AFLP) and hemolysis, elevated liver enzyme and low platelet (HELLP) syndrome. These may initially present as hypertension and proteinuria, like preeclampsia.

Like all long chain fatty acid oxidation disorders, treatment of LCHAD deficiency involves prevention of fasting and a low fat diet with provision of fats for energy as medium-chain fatty acids in the form of medium chain triglycerides. Supplementation with essential fatty acids is necessary. Prognosis is good for individuals who are diagnosed and treated rapidly and who survive early metabolic crises.

4 Diagnostic compounds

4.1 Urine organic acid profile

May show a pattern of medium chain dicarboxylic aciduria, with longer chain length 3-hydroxy-dicarboxylic aciduria present. This is a relatively non-specific finding. UOA will often be normal when the patient is metabolically stable.

4.2 Acylcarnitine profile

Free carnitine is often low, and total carnitine may be low also. Long-chain acylcarnitines, especially the hydroxy forms, C16OH-, C18OH:1- and C18OH:2-carnitines will be elevated.

4.3 Amino acids

Non-diagnostic for LCHAD deficiency.

4.4 Other important diagnostic/monitoring compounds

Glucose should be monitored routinely, along with liver enzymes.

Creatine kinase: CK will be elevated during episodic attacks in the myopathic form.

5 Newborn screening

LCHAD deficiency is one of the core disorders in the Newborn Screening RUSP. The biomarkers used to indicate LCHAD deficiency in newborn screening include low free carnitine concentrations and elevated long chain acylcarnitines species, especially C16OH-, C16:1OH-, C18:1OH- and C18:2OH-carnitines. The ratio of C16OH-carnitine to C16-carnitine (C16OH/C16) is also used.

6 Follow-up/confirmatory testing

Acylcarnitine analysis looking for elevated long chain 3-hydroxy-species is used for initial follow-up testing. The biomarkers noted above will be elevated in the newborn period and in the ill individual, but the acylcarnitine profile may be normal in the metabolically stable individual. Thus, genetic testing is usually used for confirmation of this disorder.

7 Interferences and assay or interpretation quirks

Mitochondrial FAO and carnitine transport system deficiencies which involve the long-chain fatty acid species have acylcarnitines profiles that can look alike, with elevations ranging from C14- through C18-species. This can be true also of trifunctional protein deficiency, although both TFP and LCHAD deficiency should show elevations in the long chain 3-hydroxy-acylcarnitines.

Further reading

[1] Genetics Home Reference, NIH US National Library of Medicine. https://ghr.nlm.nih.gov/condition/long-chain-3-hydroxyacyl-coa-dehydrogenase-deficiency accessed 11/4/2019.
[2] Vockley J, Longo N, Andresen BS, Bennett MJ. Mitochondrial fatty acid oxidation defects. In: Sarafoglou K, Hoffman GF, Roth KS, editors. Pediatric endocrinology and inborn errors of metabolism. 2nd ed. New York: McGraw-Hill Co; 2017. p. 125–44.
[3] NIH Genetic and Rare Diseases Information Center. LCHAD deficiency. https://rare-diseases.info.nih.gov/diseases/6867/lchad-deficiency accessed 11/4/2019.
[4] Recommended Uniform Screening Panel. HRSA. https://www.hrsa.gov/advisory-committees/heritable-disorders/rusp/index.html accessed 3/20/2019.
[5] American College of Medical Genetics Newborn Screening Expert G. Newborn screening: toward a uniform screening panel and system--executive summary. Pediatrics 2006;117:S296-307.

CHAPTER 30

Disorder: Medium-chain acyl-CoA dehydrogenase deficiency

1 Synonyms

MCAD deficiency.

2 Brief synopsis

2.1 Incidence

1:15,000

2.2 Etiology

Mitochondrial β-oxidation of fatty acids is an energy-producing pathway that is induced during fasting. Mitochondrial fatty acid β-oxidation (FAO) is a repeating four-step pathway of dehydrogenation by acyl-CoA dehydrogenases, (e.g., very long-chain acyl-CoA dehydrogenase and MCAD), hydration by enoyl-CoA hydratases, dehydrogenation by hydroxyacyl-CoA dehydrogenases, (e.g., long-chain 3-hydroxyacyl-CoA dehydrogenase), and thiolytic cleavage by ketoacyl-CoA thiolases, (e.g., beta-ketothiolase) (Fig. 30.1). At the end of each round, one molecule of acetyl-CoA is formed and can either be oxidized further in the tricarboxylic acid (TCA) cycle or converted to ketone bodies in the liver. The fatty acyl-CoA chain length is thus shortened by two carbons and re-enters the β-oxidation pathway. Many of the β-oxidation enzymes have relative affinities for different fatty acid chain length and are named accordingly (e.g., MCAD). MCAD (EC1.3.99.3) is a homotetrameric enzyme located in the mitochondrial matrix. It catalyzes the initial step in the β-oxidation of C12–C16 straight-chain fatty acyl-CoAs. A deficiency of MCAD (OMIM #201450) leads to impaired energy production from β-oxidation of these medium-chain fatty acids and impaired hepatic ketogenesis and gluconeogenesis.

MCAD deficiency is caused by mutations in the *ACADM* gene, which is located at chromosome 1p31.1. More than 80 mutations are known, with the most common mutation replacing a lysine with a glutamic acid at position 304 (p.Lys304Glu, c.985A>G). This mutation especially shows a high

A Quick Guide to Metabolic Disease Testing Interpretation
http://dx.doi.org/10.1016/B978-0-12-816926-1.00030-4

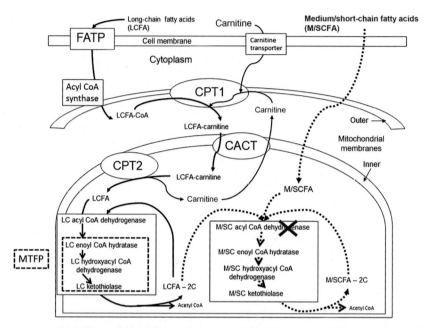

Figure 30.1 *Fatty acid oxidation pathways showing the enzyme defect in MCAD deficiency.* CACT, carnitine:acylcarnitine translocase; *CPT1*, carnitine palmitoyltransferase 1; *CPT2*, carnitine palmitoyltransferase 2; *FATP*, fatty acid transport proteins; *MTFP*, mitochondrial trifunctional protein.

carrier frequency in northern European individuals. The deficiency is inherited in an autosomal recessive manner.

3 Clinical presentation

MCAD deficiency is the most common FAO disorder and one of the most common inborn errors of metabolism. Like other fatty acid oxidation disorders, MCAD deficiency may become symptomatic in infancy due to an unrelated illness such as a viral infection leading to fasting, or other prolonged fasting such as weaning from nighttime feedings. Fasting from any cause results in fasting-associated hypoketotic hypoglycemia, vomiting, lethargy, apnea, coma, and encephalopathy. There may be growth restriction and failure to thrive. Sudden death may occur and may be mistakenly diagnosed as sudden infant death syndrome. Undiagnosed, 18 to 25% of MCAD individuals die during their first metabolic crisis. Fatty liver and fatty heart may be seen at autopsy.

Treatment involves prevention of fasting, rapid investigation of metabolic stressors, and rapid initiation of treatment to reverse catabolism and prevent hypoglycemia. Prognosis for a normal life is excellent in diagnosed individuals.

4 Diagnostic compounds

4.1 Urine organic acid profile

Urine organic acid analysis shows a characteristic pattern of elevated saturated and unsaturated dicarboxylic acids of chain length C6–C10, along with the presence of hexanoylglycine and suberylglycine.

4.2 Acylcarnitine profile

A classic pattern of elevated medium chain acylcarnitines is seen with MCAD deficiency, especially C8-, C10:1- and C10-carnitines. C6 is often also elevated. In many instances, especially when the individual is metabolically stable, the elevations are only mild to modest.

4.3 Amino acids

Non-diagnostic for MCAD deficiency.

4.4 Example chromatograph

Fig. 30.2 and Fig. 30.3.

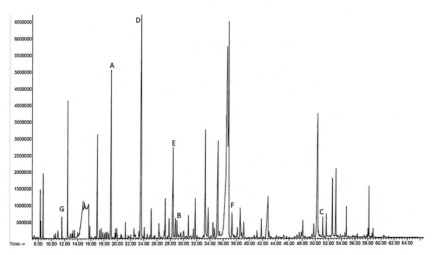

Figure 30.2 *Medium-chain acyl-CoA dehydrogenase deficiency.* (A) Internal standard (B) hexanoylglycine, (C) suberylglycine (D) adipic acid, (E) suberic acid, (F) sebacic acid with low ketones and (G) 3-hydroxybutyric acid.

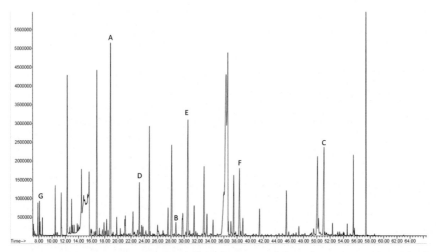

Figure 30.3 *Medium-chain acyl-CoA dehydrogenase deficiency.* (A) Internal standard (B) hexanoylglycine, (C) suberylglycine (D) adipic acid, (E) suberic acid, (F) sebacic acid with low ketones and (G) 3-hydroxybutyric acid.

4.5 Other important diagnostic/monitoring compounds

Urine acylglycine analysis may also be useful and will show the same pattern of increased medium chain acylglycines as are seen in the plasma acylcarnitine profile.

5 Newborn screening

MCAD is included on the RUSP for newborn screening.

Newborn screening for MCAD is based on tandem mass spectrometry (MS/MS) detection of elevated blood levels of the medium chain acylcarnitines, especially C8- and C10:1–carnitines.

6 Follow-up/confirmatory testing

Confirmatory testing for MCAD deficiency can be done by demonstration of two disease causing mutations by genetic analysis.

7 Interferences and assay or interpretation quirks

Urine samples collected when the individual is metabolically stable or has been treated for hypoglycemia with glucose may not show the characteristic pattern of medium chain dicarboxylic acids and hexanoyl- and suberylglycines. This pattern is most notable in patients undergoing a metabolic crisis.

Further reading

[1] GeneReviews, NIH, Medium-Chain Acyl-CoA Dehydrogenase Deficiency. https://www.ncbi.nlm.nih.gov/books/NBK1424/accessed 11/13/2019.

[2] NIH Genetic and Rare Diseases Information Center. Medium-chain acyl-CoA dehydrogenase deficiency. https://rarediseases.info.nih.gov/diseases/540/mcad-deficiency accessed 11/13/2019.

[3] Genetics Home Reference, NIH US National Library of Medicine. https://ghr.nlm.nih.gov/condition/medium-chain-acyl-coa-dehydrogenase-deficiency#   accessed 11/13/2019.

[4] Vockley J, Longo N, Andresen BS, Bennett MJ. Mitochondrial fatty acid oxidation defects. In: Sarafoglou K, Hoffman GF, Roth KS, editors. Pediatric endocrinology and inborn errors of metabolism. 2nd ed. New York: McGraw-Hill Co; 2017. p. 125–44.

CHAPTER 31

Disorder: Short chain 3-hydroxyacyl-CoA dehydrogenase deficiency

1 Synonyms

SCHAD deficiency, M/SCHAD deficiency, HADH deficiency.

2 Brief synopsis

2.1 Incidence

Unknown, extremely rare.

2.2 Etiology

Short chain 3-hydroxyacyl-CoA dehydrogenase deficiency (SCHAD deficiency, OMIM # 231530) is also known as M/SCHAD (medium/short chain) deficiency and is a disorder of the mitochondrial fatty acid β-oxidation pathway. SCHAD (EC 1.1.1.35) is the enzyme that removes a hydroxyl group from short and medium chain length fatty acids as part of the four-step fatty acid oxidation (FAO) cycle in the mitochondrial matrix. The breakdown of fatty acids is a requirement for normal functioning of high-energy requiring tissues like heart and skeletal muscle, but is also necessary for all tissues during fasting. SCHAD deficiency impairs the breakdown of fatty acids for energy (Fig. 31.1). In addition to its role in FAO, SCHAD has a "moonlighting" function in pancreatic islet cells that contributes to the clinical symptoms when SCHAD is deficient. SCHAD functions to downregulate glutamate dehydrogenase (GDH), and the absence of SCHAD increases GDH activity much like GDH gain of function mutations, causing hyperinsulinism. Thus SCHAD deficiency can be classified as a form of congenital hyperinsulinism.

The *SCHAD* gene encodes the SCHAD enzyme and is located at position 4q22-26. Although mutations in the *SCHAD* gene give rise to SCHAD deficiency and appear to be inherited in an autosomal recessive

A Quick Guide to Metabolic Disease Testing Interpretation
http://dx.doi.org/10.1016/B978-0-12-816926-1.00031-6

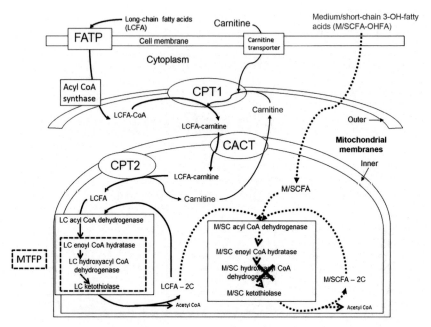

Figure 31.1 *Fatty acid oxidation pathway involving medium/short chain 3-hydroxyac-yl-CoA dehydrogenase.* *CACT*, carnitine:acylcarnitine translocase; *CPT1*, carnitine palmi-toyltransferase 1; *CPT2*, carnitine palmitoyltransferase 2; *FATP*, fatty acid transport pro-teins; *MTFP*, mitochondrial trifunctional protein.

fashion, a few patients with decreased enzyme activity have been found to have a normal gene sequence.

3 Clinical presentation

SCHAD deficiency typically presents with recurrent, episodic, hyperinsu-lin-associated hypoketotic hypoglycemia. Lethargy and seizures may oc-cur as a result of the hypoglycemia. These episodes may be precipitated by fasting, dietary protein or other stressors, and presentation generally begins within the first year of life. Unlike long–chain FAO disorders, cardiomy-opathies are not seen. Also unlike other FAO disorders, treatment involves pharmacologic interventions to control hyperinsulinism. Outcome and prognosis depends on the timing of diagnosis and appropriate treatment. Normal development is seen when treatment is timely and appropriate. Developmental delay and even death have occurred when hypoglycemic episodes are not treated quickly enough.

4 Diagnostic compounds

4.1 Urine organic acid profile

May show present or elevated concentration of 3-hydroxyglutaric acid and glutaconic acid. Glutaric acid will not be elevated. UOA may be normal when the patient is metabolically stable.

4.2 Acylcarnitine profile

Elevated 3-hydroxybutyrylcarnitine (C4—OH) is seen with SCHAD deficiency and is the most useful diagnostic biomarker for this disorder.

4.3 Amino acids

Non-diagnostic for SCHAD deficiency.

4.4 Other important diagnostic/monitoring compounds

Glucose should be monitored regularly.

Insulin: Hyperinsulinemia will be present, especially during hypoglycemic episodes.

5 Newborn screening

SCHAD deficiency is one of the disorders included in the list of secondary conditions that will be diagnosed during the differential follow-up workup of the Core conditions of the Newborn Screening RUSP. The biomarker used to indicate SCHAD deficiency in newborn screening is C4-OH-carnitine.

6 Follow-up/confirmatory testing

Acylcarnitine analysis looking for continued elevation of C4-OH-carnitine is used for initial follow-up testing, along with urine organic acids for 3-OH-glutaric acid. These biomarkers are not consistently found however, thus diagnosis relies on sequencing of the *SCHAD* gene.

7 Interferences and assay or interpretation quirks

None applicable.

Further reading

[1] Genetics Home Reference, NIH US National Library of Medicine. https://ghr.nlm.nih.gov/condition/3-hydroxyacyl-coa-dehydrogenase-deficiency# accessed 12/5/2019.

[2] Vockley J, Longo N, Andresen BS, Bennett MJ. Mitochondrial fatty acid oxidation defects. In: Sarafoglou K, Hoffman GF, Roth KS, editors. Pediatric endocrinology and inborn errors of metabolism. 2nd ed. New York: McGraw-Hill Co; 2017. p. 125–44.

[3] NIH Genetic and Rare Diseases Information Center. 3-alpha hydroxyacyl-CoA dehydrogenase deficiency. https://rarediseases.info.nih.gov/diseases/9870/mschad accessed 12/5/2019.

[4] Recommended Uniform Screening Panel. HRSA. https://www.hrsa.gov/advisory-committees/heritable-disorders/rusp/index.html accessed 3/20/2019.

[5] American College of Medical Genetics Newborn Screening Expert G. Newborn screening: toward a uniform screening panel and system--executive summary. Pediatrics 2006;117:S296-307.

CHAPTER 32

Disorder: Very long chain acyl CoA dehydrogenase deficiency

1 Synonyms

VLCAD deficiency.

2 Brief synopsis

2.1 Incidence

1:40,00–120,000

2.2 Etiology

Very long chain acyl CoA dehydrogenase deficiency (VLCAD deficiency, OMIM # 201475) is a disorder of the mitochondrial fatty acid β-oxidation pathway previously described in the chapter on MCAD deficiency. (See Chapter 30, MCAD).VLCAD (EC1.3.8.9) is the membrane bound enzyme inside the mitochondria that catalyzes the dehydrogenation of long-chain fatty acid species with chain lengths of 12–22 carbon atoms (Fig. 32.1). Like CACT and CPT2 deficiencies, when VLCAD is deficient or absent long-chain acylcarnitines build up and cannot be utilized by the mitochondrial fatty acid oxidation pathway for energy, which is especially critical during fasting or times of heavy energy expenditure. VLCAD is expressed in many tissues, including the liver, skeletal and cardiac muscle, and like CPT2, VLCAD presents as three main phenotypes.

The *ACADVL* gene encodes for VLCAD and mutations in this gene give rise to VLCAD deficiency. Its chromosomal location is 17p13.1. More than 100 mutations in the *ACADVL* gene have been described, leading to reduced or absent enzyme activity. There is a general correlation between clinical presentation and genotype with null mutations resulting in no enzyme activity causing more severe disease than mutations that result in some residual enzyme activity. VLCAD is inherited in an autosomal recessive fashion.

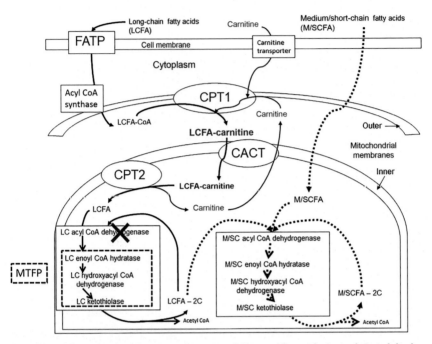

Figure 32.1 *Fatty acid oxidation pathway involving very long-chain acyl-CoA dehydrogenase.* *CACT*, carnitine:acylcarnitine translocase; *CPT1*, carnitine palmitoyltransferase 1; *CPT2*, carnitine palmitoyltransferase 2; *FATP*, fatty acid transport proteins; *MTFP*, mitochondrial trifunctional protein.

3 Clinical presentation

VLCAD deficiency has three distinct clinical presentations very similar to CPT2 deficiency. In VLCAD deficiency they seem to represent more of a continuum containing two childhood and one adult forms and presentations along that continuum depending on severity of enzyme deficiency. The most severe form is the neonatal which presents within days of birth with hypoketotic hypoglycemia and liver failure. Cardiomyopathy is present, with cardiac arrhythmias. Fasting or intercurrent infection often leads rapidly to seizures and coma. This form has a poor prognosis and may be rapidly fatal within days to months, or may become a disease of numerous recurrent episodes. Fasting tolerance is extremely low.

The later childhood onset form of VLCAD deficiency is milder, presenting with episodes of metabolic decompensation which are often triggered by fasting or illness. Cardiomyopathy is less common in this form, but other symptoms are similar to the neonatal form. As VLCAD deficient individuals get older, episodes of metabolic decompensation decrease, although many of these individuals begin to experience muscle pain upon sustained exercise, like the adult onset form of VLCAD deficiency.

The adult-onset, myopathic form of VLCAD deficiency presents with recurrent episodes of muscle pain, rhabdomyolysis and myoglobinuria which can be triggered by exercise, fasting, exposure to cold, and various forms of stress. Myoglobinuria accompanying these attacks can lead to renal failure. Creatine kinase concentrations will be massively elevated during attacks but are usually normal between attacks.

Like other long-chain FAO disorders, treatment of VLCAD deficiency involves prevention of fasting and a low fat diet with provision of fats for energy as medium-chain fatty acids in the form of medium chain triglycerides. In the case of the myopathic form, avoidance of triggers is important.

4 Diagnostic compounds

4.1 Urine organic acid profile

May show a pattern of medium chain dicarboxylic aciduria. This is a non-specific finding. UOA will often be normal when the patient is metabolically stable.

4.2 Acylcarnitine profile

Free carnitine is often low, and total carnitine may be low also. Long-chain acylcarnitines, especially C14:1- and C14- carnitines will be elevated, especially early in the neonatal period, but may be completely normal between decompensation episodes.

4.3 Amino acids

Non-diagnostic for VLCAD deficiency.

4.4 Other important diagnostic/monitoring compounds

Ammonia: Hyperammonemia is important to monitor in all FAO disorders.

Creatine kinase: CK will be elevated during attacks in the myopathic form and may be elevated at times in the severe neonatal form.

5 Newborn screening

VLCAD deficiency is one of the core disorders in the Newborn Screening Recommended Uniform Screening Panel (RUSP). The biomarkers used to indicate VLCAD deficiency in newborn screening include low free carnitine concentrations and elevated long chain acylcarnitines species, especially C14:1- and C14-carnitines. Because false positive screening results are common for VLCAD (due to low thresholds set my most programs to optimize sensitivity), some states have incorporated second tier mutational analysis to check for the most common variants.

6 Follow-up/confirmatory testing

Because C14- and C14:1-carnitines may not be abnormal on follow-up testing for VLCAD deficiency, genetic testing for mutations in the *ACAD-VL* gene is necessary for confirmation of this disorder.

7 Interferences and assay or interpretation quirks

All mitochondrial FAO disorders which involve the long-chain fatty acid species have acylcarnitines profiles that can look alike, with elevations ranging from C14- through C18-species. Looking for the predominant elevations, in this case C14:1- and C14-carnitines, can help interpret to profile, as can looking at the clinical picture and other lab tests results.

Further reading

[1] Genetics Home Reference, NIH US National Library of Medicine. https://ghr.nlm.nih.gov/condition/very-long-chain-acyl-coa-dehydrogenase-deficiency#   accessed 11/25/2019.
[2] Vockley J, Longo N, Andresen BS, Bennett MJ. Mitochondrial fatty acid oxidation defects. In: Sarafoglou K, Hoffman GF, Roth KS, editors. Pediatric endocrinology and inborn errors of metabolism. 2nd ed. New York: McGraw-Hill Co; 2017. p. 125–44.
[3] GeneReviews, NIH, Very Long-Chain Acyl-CoA Dehydrogenase Deficiency. https://www.ncbi.nlm.nih.gov/books/NBK6816/accessed 11/26/2019.
[4] NIH Genetic and Rare Diseases Information Center. VLCAD Deficiency. https://rarediseases.info.nih.gov/diseases/5508/vlcad-deficiency accessed 12/6/2019.

Other metabolic disorders

Disorder: Biotin: Biotinidase deficiency and holocarboxylase synthetase deficiency

1 Synonyms

Multiple carboxylase deficiency.

2 Brief synopsis

2.1 Incidence

Biotinidase: ~1:60,000–80,000 newborns.
Holocarboxylase synthetase: < 1:200,000 newborns.

2.2 Etiology

Multiple carboxylase deficiency is caused by either biotinidase deficiency (BTD, OMIM # 253260) or holocarboxylase synthetase deficiency (HLC-SD, OMIM # 253270). Both these enzymes deliver biotin to multiple biotin-dependent carboxylases. Both enzymes are required for the recycling and re-utilization of biotin (Fig. 33.1) and the proper functioning of the biotin-dependent carboxylases (Fig. 33.2). Biotinidase (EC3.5.1.12) is the enzyme that frees biotin from its protein conjugates to allow it to be recycled for use by the biotin-requiring carboxylases. Holocarboxylase synthetase (EC6.3.4.10), also known as biotin-[propionyl-CoA-carboxylase (ATP hydrolyzing)] synthetase or biotin-[propionyl-CoA-carboxylase (ATP hydrolyzing)]ligase, is the enzyme that catalyzes the covalent binding of biotin with the four apocarboxylases to form the active carboxylases. These carboxylases are involved in gluconeogenesis, fatty acid synthesis, and several catalytic pathways for amino acids and odd-chain fatty acids.

Deficiencies in biotinidase or holocarboxylase synthetase may result in elevated concentrations of multiple substrates of the biotin-requiring carboxylases. Additionally, free biotin concentrations may be low in these disorders. (Figs. 33.1 and 33.2). The gene locus for biotinidase is 3p25.1 and for holocarboxylase synthetase it is 21q22.13. Both disorders are inherited in an

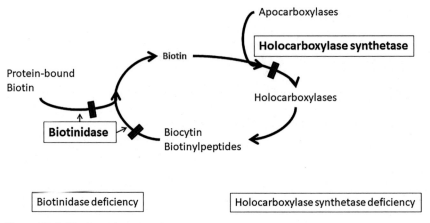

Figure 33.1 *Biotin pathway.*

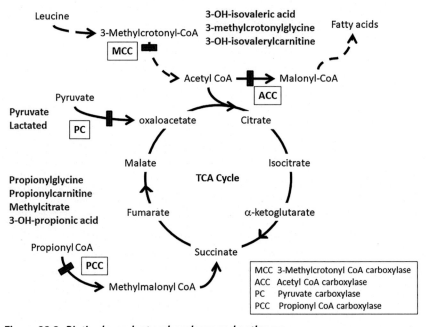

Figure 33.2 *Biotin-dependent carboxylases and pathways.*

autosomal recessive fashion. Pathogenic variants of the biotinidase gene can be found throughout the coding sequence of the gene, with more the 150 reported mutations causing biotinidase deficiency. About 35 mutations have been reported for the holocarboxylase synthetase gene, with the majority of them being found in the area of HCS believed to be responsible for biotin binding.

3 Clinical presentation

Biotinidase deficiency may present between 1 week and 10 years of age, with the mean age of presentation being 3–6 months of age. Profound deficiency may result in seizures, hypotonia, movement and balance problems, breathing problems, hearing and vision loss, skin rashes and hair loss and global developmental delay. Milder deficiencies may present with hypotonia, skin rashes and hair loss that appear only with stress or intercurrent infections and illnesses. Treatment involves biotin supplementation and most cases respond well to biotin treatment.

Holocarboxylase synthestase deficiency may present anywhere from shortly after birth to about 6 years of age, with roughly half of patients presenting acutely in the newborn period. Symptoms of acute presentation include severe metabolic acidosis, lethargy, hypotonia, vomiting, seizures, hypothermia and unconsciousness, which can progress quickly to coma and death. If biotin treatment is started before neurological damage occurs, resolution of clinical symptoms generally occurs.

4 Diagnostic compounds

4.1 Urine organic acid profile

Results are not consistent between patients, but may show accumulations of metabolites upstream from the enzymes which require biotin for function. Thus, the metabolites seen on urine organic acid profile may include 3-hydroxyisovalerate, 3-hydroxypropionate, methylcitrate, tiglylglycine, propionylglycine and 3-methylcrotonylglycine.

4.2 Acylcarnitine profile

Acylcarnitine profiles will also show inconsistent metabolites from the biotin-requiring enzymes, including propionylcarnitine (C3-carnitine) and 3-hydroxyisovalerylcarnitine (C5OH-carnitine). Low free carnitine is seen in plasma.

4.3 Amino acids

Non-diagnostic for biotin enzyme deficiencies.

4.4 Other important diagnostic/monitoring compounds

Lactic acid is generally elevated in both plasma and in the urine organic acid. Hyperammonemia is observed with HLCSD.

5 Newborn screening

Biotinidase deficiency and holocarboxylase synthetase deficiency are both included in the newborn screening programs of all 50 States. Elevated C5OH-carnitine is used to screen for these deficiencies, but is not specific for them. C3-carnitine may also be elevated.

6 Follow-up/confirmatory testing

Diagnosis of BTD can be made by measuring serum biotinidase activity (colorimetric measurement). Molecular testing for pathogenic mutations in the BTD or HCS is available.

7 Interferences and assay or interpretation quirks

Serum and urine biochemical markers may not be present in individuals with disorders of biotin metabolism, depending on the sufficiency of biotin present. The diagnosis of these disorders will depend on measurement of enzyme activity or molecular testing.

Further reading

[1] GeneReviews, NIH, Biotinidase deficiency. https://www.ncbi.nlm.nih.gov/books/NBK1322/accessed 11/01/2018.
[2] Genetics Home Reference, NIH US National Library of Medicine. https://ghr.nlm.nih.gov/condition/biotinidase-deficiency#definition accessed 11/01/2018.
[3] Hoffmann GF, Burlina A, Barshop BA. Organic acidurias. In: Sarafoglou K, Hoffman GF, Roth KS, editors. Pediatric endocrinology and inborn errors of metabolism. 2nd ed. New York: McGraw-Hill Co; 2017. p. 209–50.
[4] Pasquali M, De Biase, I. Newborn Screening. In: Jones PM, Haymond S, Dietzen DJ, Bennett MJ, Eds. Pediatric Laboratory Medicine. 2017. McGraw-Hill Co. New York. 95–134.

CHAPTER 34

Disorder: Canavan Disease

1 Synonyms

aspartoacylase deficiency, aminoacylase 2 deficiency, *N*-acetylaspartic acid-uria, ASPA Deficiency

2 Brief Synopsis

2.1 Incidence

~1:200,000 newborns; 1:5,000-14,000 in Ashkenazi Jewish population

2.2 Etiology

Aspartoacylase deficiency (OMIM # 271900) is a neurodegenerative disorder first described by Dr Myrtelle Canavan in 1931. It is a type of leukodystrophy.

The deficient enzyme, aspartoacylase (EC3.5.1.15), also known as aminoacylase II, deacetylates *N*-acetyl aspartic acid (NAA), which is proposed to be important for myelin synthesis and water transport in the neurons. NAA is found in high concentrations in the normal brain, second only to glutamate. Deficiency in aspartoacylase results in accumulation of NAA in the brain (Fig. 34.1), causing water retention in oligodenrocytes and lack of myelination as well as progressive demyelination.

The *ASPA* gene locus for this enzyme is 17p13.2 and the disorder is inherited in an autosomal recessive fashion. Two common mutations in the *ASPA* gene are found to be the cause in 99% of the cases in the Ashkenazi Jewish population, p.Glu285Ala and pTyr231Ter. In the non-Ashkenazi Jewish population, a different mutation, pAla305Glu, is the most common, causing 30-60% of the cases.

Disease outcome and prognosis are dependent on the degree of aspartoacylase deficiency present.

3 Clinical Presentation

Individuals with Canavan's disease have impaired neuronal activity in the brain apparently due to disrupted growth and/or maintenance of myelin

A Quick Guide to Metabolic Disease Testing Interpretation
http://dx.doi.org/10.1016/B978-0-12-816926-1.00034-1

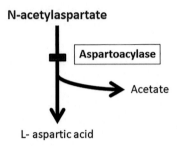

N-acetylaspartate

Aspartoacylase

Acetate

L- aspartic acid

Figure 34.1 *Pathway involving aspartoacylase.*

sheaths, which protect neurons and help transmit nerve impulses. The most common presentation is the neonatal/infantile form, which presents between 2–5 months of age and is also the most severe form of the disorder. Common symptoms include hypotonia, macrocephaly, irritability, feeding problems, seizures and sleep disturbances. Motor skills do not develop and often regression of early motor development occurs. The infant is not capable of turning over, controlling the head or sitting without support. There is currently no treatment for this disorder and these children in general do not live into adolescence as the often rapid, progressive neurodegeneration is ultimately fatal.

The mild/juvenile form of Canavan's disease is less common and presents with mild developmental delay of speech and motor skills. Prognosis is good in these children and they may only require speech therapy and tutoring. Lifespan is not shortened.

4 Diagnostic compounds

4.1 Urine organic acid profile

N-acetylaspartic acid is detected in the urine sample. This compound may also be found in the CSF and plasma.

4.2 Acylcarnitine profile

Non-diagnostic for Canavan's disease

4.3 Amino acids

Non-diagnostic for Canavan's disease

4.4 Example chromatograph

Fig. 34.2.

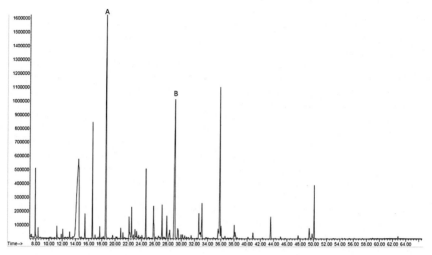

Figure 34.2 *Canavan Disease.* (A) Internal standard and (B) *n*-acetylaspartic acid.

4.5 Other important diagnostic/monitoring compounds

None applicable.

5 Newborn screening

Canavan's disease is not included in newborn screening programs as it is not detected by methods used for newborn screening.

6 Follow-up/confirmatory testing

Elevated excretion of *N*-acteylaspartate on urine organic acid analysis is generally considered diagnostic for this condition. Full gene sequencing or mutational analysis of the *ASPA* gene is available to aid in diagnosis.

7 Interferences and assay or interpretation quirks

Occasionally a low concentration of *N*-acetylaspartate is detected in a urine organic acid analysis, which is not diagnostic and not confirmed upon repeat analysis. As always, results should be correlated with clinical picture and questionable results should be repeated.

Further reading

[1] Genetics Home Reference, NIH US National Library of Medicine. Canavan Disease https://ghr.nlm.nih.gov/condition/canavan-disease#definition accessed 3/21/2019.

[2] GeneReviews, NIH, Canavan Disease. https://www.ncbi.nlm.nih.gov/books/NBK1234/accessed 3/21/2019.

[3] Hoffmann GF, Burlina A, Barshop BA. Organic Acidurias. In: Sarafoglou K, Hoffman GF, Roth KS, editors. Pediatric Endocrinology and Inborn Errors of Metabolism. 2nd ed. New York: McGraw-Hill Co; 2017. p. 209–50.

CHAPTER 35

Disorder: Dihydropyrimidine dehydrogenase deficiency

1 Synonyms

Dihydropyrimidinuria, familial pyrimidinuria, hereditary thymine-uraciluria, DPD deficiency, DPYD deficiency.

2 Brief synopsis

2.1 Incidence

Unknown, rare.

2.2 Etiology

Dihydropyrimidine dehydrogenase deficiency (DPD deficiency, OMIM # 274270) is a defect in pyrimidine catabolism. Dihydropyrimidine dehydrogenase (EC1.3.1.2) is the enzyme that begins the catabolism of the pyrimidines, thymine and uracil, by adding hydrogen and converting them to dihydrothymine and dihydrouracil respectively. Deficiency in this enzyme causes elevated concentrations of these two pyrimidines to accumulate in the body and be excreted in the urine (Fig. 35.1). It is not clear how accumulations of the pyrimidines lead to the specific symptoms seen with this deficiency. Additionally, there is no genotype-phenotype correlation; however, the symptoms appear to be more severe in cases of complete enzyme deficiency than in partial deficiency cases.

DPD deficiency is inherited in an autosomal recessive manner. The gene that codes for dihydropyrimidine dehydrogenase is the *DPYD* gene and its location is 1p21.3. There are over 50 known mutations in the *DYPD* gene that result in DPD deficiency.

3 Clinical presentation

In cases of severe DPD deficiency, presentation is in infancy with neurological problems, including recurrent seizures and developmental delay. Other symptoms that are sometimes seen include microcephaly, autistic

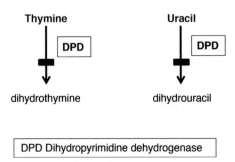

Figure 35.1 *Pathway involving dihydropyrimidine dehydrogenase.*

behaviors, intellectual disability and increased muscle tone. Less severe cases with partial enzyme activities may be asymptomatic. All individuals with DPD deficiency, asymptomatic or not, are vulnerable to possibly life-threatening toxic reactions to fluoropyrimidine drugs such as 5-fluorouracil (5-FU), since these drugs cannot be broken down efficiently. As a result, DPD is also classified as a pharmacogenetic disorder. Symptoms of severe toxicity in patients with DPD include GI symptoms such as abdominal pain, nausea, vomiting and diarrhea due to ulceration of the GI tract, as well as neutropenia, thrombocytopenia and hemorrhage. In some cases, toxic reactions can extend to severe neurological sequelae with loss of speech and inability to move extremities. Because of the severity of reactions, testing for DPD deficiency may be suggested for individuals prior to starting treatment with 5-fluorouracil. Even though the disorder is rare, it has been estimated that 2 – 8% of the population may be vulnerable to toxic reactions to 5-FU.

There are no defined treatment modalities for DPD deficiency. Seizures are treated with anticonvulsants.

4 Diagnostic compounds

4.1 Urine organic acid profile

Organic acid analysis will show elevated excretion of uracil and/or thymine.

4.2 Acylcarnitine profile

Non-diagnostic for DPD deficiency.

4.3 Amino acids

Non-diagnostic for DPD deficiency

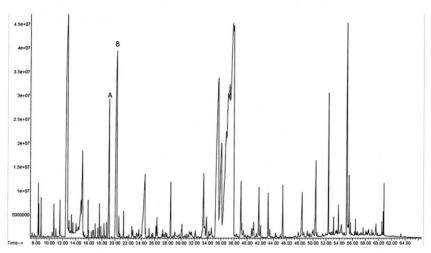

Figure 35.2 *Dihyropyrimidine dehydrogenase deficiency.* (A) Internal standard and (B) thymine.

4.4 Example chromatograph

Fig. 35.2.

4.5 Other important diagnostic/monitoring compounds

None applicable.

5 Newborn screening

DPD deficiency is not included in the RUSP of newborn screening programs, not as a core condition nor as a secondary condition.

6 Follow-up/confirmatory testing

Although the presence of thymine and uracil in an organic acid profile is suggestive of DPD deficiency, diagnosis should be confirmed using genetic testing.

7 Interferences and assay or interpretation quirks

Thymine and uracil in the urine may also be detected in cases of dihydropyrimidinase deficiency which is the next step in the pathway of pyrimidine catabolism after DPD. Uracil may also be detected in some urea cycle disorders (OTC deficiency and citrullinemia). Both thymine and uracil can be present in the urine from metabolism of caffeine.

Further reading

[1] Genetics Home Reference, NIH US National Library of Medicine. https://ghr.nlm.nih.gov/condition/dihydropyrimidine-dehydrogenase-deficiency# accessed 10/28/2019.
[2] Nyhan WL. Purine and pyrimidine metabolism. In: Sarafoglou K, Hoffman GF, Roth KS, editors. Pediatric endocrinology and inborn errors of metabolism. 2nd ed. New York: McGraw-Hill Co; 2017. p. 1023–54.

CHAPTER 36

Disorder: Glutathione synthetase deficiency

1 Synonyms

Pyroglutamic aciduria, pyroglutamic acidemia; 5-oxoprolinuria, 5-oxoprolinemia.

2 Brief synopsis

2.1 Incidence

Unknown, rare.

2.2 Etiology

Glutathione synthetase deficiency (GSSD, OMIM #266130) is caused by mutations of the gene that encodes the glutathione synthetase enzyme (*EC6.3.2.3*). Glutathione synthetase catalyzes the synthesis of glutathione (γ-glutamyl-cysteinylglycine) by conjugation of γ-glutamylcysteine and glycine as part of the γ-glutamyl cycle (Fig. 36.1). Absence or deficiency in this enzyme disrupts the γ-glutamyl cycle and results in a lack of glutathione and over-production of pyroglutamic acid, also known as 5-oxoproline. Glutathione acts as an antioxidant, protecting cells from reactive oxygen species produced during normal metabolism, as well as playing a role in the metabolism of drugs and carcinogens. Deficiency of glutathione results in the symptoms associated with this disorder.

The gene that encodes glutathione synthetase is the *GSS* gene located on chromosome 20q11.2. More than 30 mutations in the *GSS* gene have been identified in GSSD, and the disorder is inherited in an autosomal-recessive manner.

3 Clinical presentation

Glutathione synthetase deficiency may present as mild, moderate or severe disease. The mild form generally presents only as a hemolytic anemia with pyroglutamic acid present in the urine on organic acid analysis. Moderate

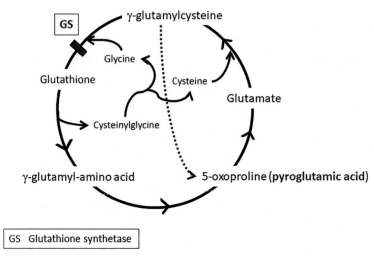

Figure 36.1 *Pathway involving glutathione synthetase.*

GSSD usually presents not long after birth with hemolytic anemia, pyroglutamic aciduria and metabolic acidosis. Severe GSSD will have these symptoms plus seizures, psychomotor retardation and ataxia, and intellectual disabilities. Recurrent bacterial infections also tend to accompany the severe form of GSSD.

There is no defined treatment for GSSD, but instead involves managing the medical problems associated with the disorder. Therefore, the severe forms of GSSD have a poor prognosis.

4 Diagnostic compounds

4.1 Urine organic acid profile

Organic acid profile will show a large peak of pyroglutamic acid. Lactic acid may also be excreted in elevated amounts.

4.2 Acylcarnitine profile

Non-diagnostic for GSSD.

4.3 Amino acids

Non-diagnostic for GSSD.

4.4 Example chromatograph

Fig. 36.2 and Fig. 36.3.

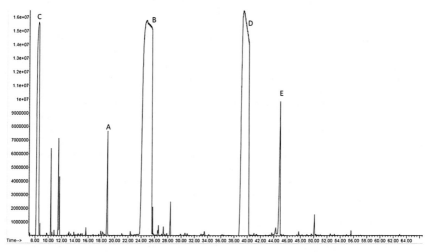

Figure 36.2 *Glutathione synthase deficiency.* (A) Internal standard (B) pyroglutamic acid, (C) lactic acid, (D) 4-hydroxyphenyllactic acid and (E) 4-hydroxyphenylpyruvic acid.

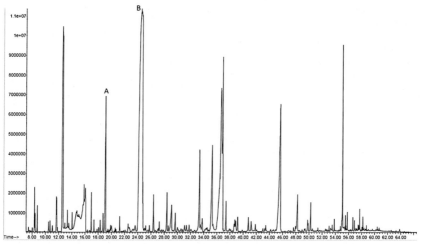

Figure 36.3 *Glutathione synthase deficiency.* (A) Internal standard and (B) pyroglutamic acid.

4.5 Other important diagnostic/monitoring compounds

None applicable.

5 Newborn screening

GSSD is not included in newborn screening programs.

6 Follow-up/confirmatory testing

Urine organic acid analysis is currently the only test clinically available for the diagnosis of GSSD.

7 Interferences and assay or interpretation quirks

Pyroglutamic acid can be present in the urine from exogenous sources (endogenous or exogenous glutamine degradation, acetaminophen, vigabatrin, fludoxacillin).

Further reading

[1] Genetics Home Reference, NIH US National Library of Medicine. https://ghr.nlm.nih.gov/condition/glutathione-synthetase-deficiency#definition accessed 10/30/2019.
[2] NIH Genetic and Rare Diseases Information Center. Glutathione Synthetase deficiency. https://rarediseases.info.nih.gov/diseases/10047/glutathione-synthetase-deficiency accessed 10/30/2019.

CHAPTER 37

Disorder: Pyruvate dehydrogenase deficiency

1 Synonyms

PDH deficiency, pyruvate dehydrogenase complex deficiency.

2 Brief Synopsis

2.1 Incidence

Unknown, rare.

2.2 Etiology

Pyruvate dehydrogenase deficiency (PDH deficiency, OMIM # 312170, 614111, 245348, 245349) is a disorder of pyruvate metabolism caused by multiple defects in the pyruvate dehydrogenase multienzyme complex. Pyruvate sits at a highly regulated crossroads of metabolism and can be further metabolized in any of four directions, depending on the metabolic needs of the body. Pyruvate can be converted to lactate by lactate dehydrogenase, to alanine by alanine transaminase, to oxaloacetate by pyruvate carboxylase, or to acetyl CoA by the pyruvate dehydrogenase complex. Deficiency in PDH results in a massive back-up to lactate (Fig. 37.1) and a deficiency in cellular energy production, which cause life-threatening sequelae. Pyruvate dehydrogenase (EC1.2.4.1) is an enzyme in the pyruvate dehydrogenase multienzyme complex, which has subunits in common with both the 2-ketoglutarate dehydrogenase and the branch chain α-ketoacid dehydrogenase multienzyme complexes. Like these two other complexes PDH multienzyme complex is comprised of three catalytic enzymes and two regulatory enzymes that control the complex activities. Defects in this complex enzyme system lead to PDH deficiency.

The most common cause of PDH deficiency are mutations in the *PDHA1* gene which accounts for 60%–80% of cases. This gene codes for the E1 alpha subunit of pyruvate dehydrogenase. Its chromosomal location is Xp22.12 and it is inherited in an X-linked fashion. Mutations in *PDHB*

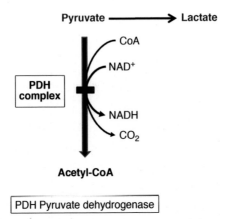

Figure 37.1 *Pathway showing the enzyme defect in PDH deficiency.*

(coding for E1 beta subunit), *DLAT* (coding for E2 enzyme), *PDHX* (coding for E3 protein) and *PDP1* (coding for the pyruvate dehydrogenase phosphatase regulatory enzyme) have all been seen in individuals with PDH deficiency. PDH deficiency caused by these other mutations is inherited in an autosomal recessive manner.

3 Clinical presentation

Pyruvate dehydrogenase deficiency demonstrates a wide range of presentations from fatal congenital lactic acidosis with brain malformations to a later age presentation of mild ataxia and some neuropathy with normal cognitive function. Most cases of PDH deficiency present like other mitochondrial disorders with severe lactic acidosis, brain abnormalities, hypotonia, feeding difficulties, seizures and Leigh's syndrome of rapidly progressive encephalopathy with some constellation of symptoms including optic atrophy, opthalmoplegia, psychomotor regression, dystonia, ataxia, and central respiratory difficulties.

There is no proven treatment for PDH deficiency, however, empirically it makes sense that replacing carbohydrate with a ketogenic diet should bypass the metabolic block. Avoiding carbohydrates and supplementing ketones is the usual treatment, although those individuals with severe neurological impairment do not appear to benefit. Outcomes of

PDH deficiency range from early death to survival with varying amounts of cognitive defects.

4 Diagnostic compounds

4.1 Urine organic acid profile

Urine organic acid analysis shows a pattern of lactic acid excretion.

4.2 Acylcarnitine profile

Non-diagnostic for PDH.

4.3 Amino acids

Alanine is frequently elevated, especially during times of high lactic acidosis.

4.4 Other important diagnostic/monitoring compounds

CSF lactate and pyruvate measurement can be useful if the blood lactate is normal.

5 Newborn screening

PDH is not included as a core condition on the RUSP, nor as a secondary condition for newborn screening, and is not screened for.

6 Follow-up/confirmatory testing

Elevated plasma concentrations of both lactate and pyruvate with a normal lactate to pyruvate (L:P) ratio is suggestive of this disorder along with an elevated alanine, however biochemical markers are not diagnostic. Confirmatory tests rely on proven pathogenic mutations upon genetic testing. Because mutations in *PDHA* are *X*-linked and lyonization occurs, females may or may not show reduced enzyme activity, and have more variability in symptoms than males.

7 Interferences and assay or interpretation quirks

None applicable.

Further reading

[1] NIH Genetic and Rare Diseases Information Center. Pyruvate Dehydrogenase Deficiency.   https://rarediseases.info.nih.gov/diseases/7513/pyruvate-dehydrogenase-deficiency accessed 12/13/2019.
[2] Genetics Home Reference, NIH US National Library of Medicine. https://ghr.nlm.nih.gov/condition/pyruvate-dehydrogenase-deficiency# accessed 12/13/2019.
[3] Kerr DS, Bedoyan JK. Disorders of pyruvate metabolism and the tricarboxylic acid cycle. In: Sarafoglou K, Hoffman GF, Roth KS, editors. Pediatric endocrinology and inborn errors of metabolism. 2nd ed. New York: McGraw-Hill Co; 2017. p. 105–24.

Index

Note: Page numbers followed by "f" indicate figures, "t" indicate tables.

Printed in the United States
By Bookmasters